JUPITER: Joint Universal Parameter IdenTification and Evaluation of Reliability — An Application Programming Interface (API) for Model Analysis

Edited by Edward R. Banta,[1] Eileen P. Poeter,[2] John E. Doherty,[3] and Mary C. Hill[1]

Prepared in cooperation with the
U.S. Environmental Protection Agency

Techniques and Methods 6-E1

U.S. Department of the Interior
U.S. Geological Survey

Denver, Colorado
2006

[1]U.S. Geological Survey, Lakewood, Colorado and Boulder, Colorado, USA

[2]Colorado School of Mines and International Ground Water Modeling Center, Golden, Colorado, USA

[3]University of Queensland, Brisbane, and Watermark Numerical Computing, Corinda, Queensland, Australia

U.S. DEPARTMENT OF THE INTERIOR

DIRK KEMPTHORNE, Secretary

U.S. GEOLOGICAL SURVEY

Mark Myers, Director

For additional information write to: Copies of this report can be purchased from:

Colorado Water Science Center   U.S. Geological Survey
U.S. Geological Survey   Branch of Information Services
Box 25046, MS 415   Box 25286
Denver Federal Center   Denver, CO 80225-0425
Denver, CO 80225

Preface

This report describes the capabilities and use of the JUPITER API — the Joint Universal Parameter IdenTification and Evaluation of Reliability Application Programming Interface. This API provides resources for programming applications (computer programs) designed to analyze process models. Process models simulate, for example, physical, chemical, and(or) biological processes of a site or experiment of interest. The analyses involved could include sensitivity analysis, data needs assessment, calibration, uncertainty evaluation, optimization, and other types of applications that require similar capabilities.

The documentation presented in this report includes descriptions of the conventions, data, subroutines, functions, and generic interfaces that constitute the API. Information on most of the statistical methods included in the API are presented by Hill and Tiedeman (in press) and references cited therein.

The performance of the JUPITER API has been tested in a variety of applications. Future use, however, might reveal errors that were not detected in the test simulations. Users are requested to notify the originating office of any errors found in the report or the API. Updates might occasionally be made to both the report and to the API. Users can check for updates on the Internet at URL:

http://water.usgs.gov/nrp/gwsoftware/jupiter/jupiter_api.html

Contents

Tables

Figures

SYMBOLS AND ABBREVIATIONS

Symbol or Abbreviation	Meaning
AIC	Akaike information criterion. Used to discriminate between alternative models.
AICc	Corrected Akaike information criterion. Used to discriminate between alternative models.
BIC	Bayesian information criterion. Used to discriminate between alternative models.
CSS	Composite scaled sensitivity. Used in sensitivity analysis.
DSS	Dimensionless scaled sensitivity. Used in sensitivity analysis.
HQ	Hannan and Quinn's Criterion. Used to discriminate between alternative models.
k	Number of estimated parameters + 1
KIC	Kashyap's Criterion. Used to discriminate between alternative models.
MLOF	Maximum likelihood objective function
MPR	Number of prior-information equations
n	Number of observations used + number of parameter-information equations used
NOBS	Number of observations
NPE	Number of adjustable parameters
NPERD	Number of estimated parameters
R2N	Correlation coefficient for standard standard normal statistic and weighted residuals. Used to test independence and normality of weighted residuals.
SSWR	Sum of squared, weighted residuals
X	Sensitivity matrix
X^T	Transpose of sensitivity matrix
σ^2	Estimated error variance
ω	Weight matrix

JUPITER: Joint Universal Parameter IdenTification and Evaluation of Reliability — An Application Programming Interface (API) for Model Analysis

Edited by Edward R. Banta, Eileen P. Poeter, John E. Doherty, and Mary C. Hill

Abstract

The **J**oint **U**niversal **P**arameter **I**den**T**ification and **E**valuation of **R**eliability Application **P**rogramming **I**nterface (JUPITER API) improves the computer programming resources available to those developing applications (computer programs) for model analysis.

The JUPITER API consists of eleven Fortran-90 modules that provide for encapsulation of data and operations on that data. Each module contains one or more entities: data, data types, subroutines, functions, and generic interfaces. The modules do not constitute computer programs themselves; instead, they are used to construct computer programs. Such computer programs are called applications of the API. The API provides common modeling operations for use by a variety of computer applications.

The models being analyzed are referred to here as process models, and may, for example, represent the physics, chemistry, and(or) biology of a field or laboratory system. Process models commonly are constructed using published models such as MODFLOW (Harbaugh et al., 2000; Harbaugh, 2005), MT3DMS (Zheng and Wang, 1996), HSPF (Bicknell et al., 1997), PRMS (Leavesley and Stannard, 1995), and many others. The process model may be accessed by a JUPITER API application as an external program, or it may be implemented as a subroutine within a JUPITER API application . In either case, execution of the model takes place in a framework designed by the application programmer. This framework can be designed to take advantage of any parallel processing capabilities possessed by the process model, as well as the parallel-processing capabilities of the JUPITER API.

Model analyses for which the JUPITER API could be useful include, for example:

1. Compare model results to observed values to determine how well the model reproduces system processes and characteristics.

2. Use sensitivity analysis to determine the information provided by observations to parameters and predictions of interest.

3. Determine the additional data needed to improve selected model predictions.

4. Use calibration methods to modify parameter values and other aspects of the model.

5. Compare predictions to regulatory limits.

6. Quantify the uncertainty of predictions based on the results of one or many simulations using inferential or Monte Carlo methods.

7. Determine how to manage the system to achieve stated objectives.

The capabilities provided by the JUPITER API include, for example, communication with process models, parallel computations, compressed storage of matrices, and flexible input capabilities. The input capabilities use input blocks suitable for lists or arrays of data. The input blocks needed for one application can be included within one data file or distributed among many

1

files. Data exchange between different JUPITER API applications or between applications and other programs is supported by data-exchange files.

The JUPITER API has already been used to construct a number of applications. Three simple example applications are presented in this report. More complicated applications include the universal inverse code UCODE_2005 (Poeter et al., 2005), the multi-model analysis MMA (Eileen P. Poeter, Mary C. Hill, E.R. Banta, S.W. Mehl, and Steen Christensen, written commun., 2006), and a code named OPR_PPR (Matthew J. Tonkin, Claire R. Tiedeman, mary C. Hill, and D. Matthew Ely, written communication, 2006).

This report describes a set of underlying organizational concepts and complete specifics about the JUPITER API. While understanding the organizational concept presented is useful to understanding the modules, other organizational concepts can be used in applications constructed using the JUPITER API.

Chapter 1: Overview of JUPITER API
By Edward R. Banta and Mary C. Hill

Building models that represent geologic, hydrogeologic, and other natural systems requires that data be assimilated into the modeling process in a meaningful way. Components of effective assimilation include, for example,

1. quantify data error and properly account for it during data assimilation,
2. evaluate the information provided by existing data to parameters and predictions of interest as needed to design a model that is both tractable and useful,
3. identify new data that are most likely to improve model predictions,
4. discriminate between alternative models, and
5. determine parameter values and distributions that produce the best fit between simulated values and associated data and compare the values and distribution against other kinds of data.

Once developed, models are often used to quantify prediction uncertainty and to manage the system to achieve stated objectives.

Thus, development and use of models require a range of methods, including sensitivity analysis, data needs assessment, inverse problems, uncertainty evaluation, and optimization. These problems have been addressed by many researchers in many fields. Recent textbooks include, for example, Tarantola (1994), Parker (1994,) Menke (1984), and Saltelli and others (2000, 2004) from the geophysics community; Sun (1994), Ahlfeld and Mulligan (2000), and Hill and Tiedeman (in press) from the ground-water community; Burnham and Anderson (2002) from the biological community, and Cook and Weisberg (1982, 1999), Dennis and Schnabel (1996), and Draper and Smith (1998) from the statistics community. Despite the substantial investment in this field, the problems are not solved. The problems are difficult because natural systems are complex, available data do not fully characterize the systems, computers are only now becoming adequate to address many of the relevant issues, and societal demands are increasing on many systems of concern.

In this difficult environment, advances can be made more rapidly if researchers can more readily test new ideas on realistic problems. There are some resources to accomplish this, such as The DAKOTA toolkit (http://endo.sandia.gov/DAKOTA/software.html) and the COSU API (Babendreier, 2004). However, all existing resources have significant deficiencies. The enhancements provided by the JUPITER API to address these deficiencies are

(a) comprehensive methods for interacting with the complex process models commonly needed to represent the systems of concern (the only requirements of the process models are that they have text-only input and output files and can be invoked by an operating-system command);
(b) improved communication methods between applications and other computer programs;
(c) a robust method for parallel computations (the only requirement is network read and write access between computers);
(d) a flexible, largely self-descriptive input design that facilitates reuse of constructed input in different applications;
(e) inclusion of a useful set of statistic and sensitivity-analysis techniques, some of which are not readily available thorough other resources;
(f) more intuitive methods for accounting for data error; and
(g) compressed storage of matrices (as needed for typically large, sparse weight matrices that commonly embody data error).

3

Existing programs, such as UCODE (Poeter and Hill, 1998) and PEST (Doherty, 2004) have many of the missing capabilities, but their structure is not designed to support development of alternative ideas by many application developers.

The JUPITER API consists of eleven Fortran-90 modules that provide for encapsulation of data and operations on that data. Each module is comprised of some combination of data, derived data types, subroutines, functions, and generic interfaces. The modules do not constitute computer programs themselves; instead, they are used to construct computer programs. Such computer programs are called applications of the API, and could include capabilities from other systems such as the COSU API or DAKOTA.

The JUPITER API makes commonly needed capabilities readily accessible to a variety of applications through its modular construction. Modularity is attained in large part by adhering to structured-programming concepts and by using Fortran-90 modules. Data and subroutines, functions, and generic interfaces that are closely related to a common purpose generally are located in a module designed for that purpose. So, for example, there is a Sensitivity Module, a Statistics Module, and a Parallel-Processing Module. Subroutines, functions, and generic interfaces that are used for many purposes are located in a Utilities Module. Data needed by many modules are in a Global Data Module.

The remaining chapters of this document describe the underlying organizational concepts and the details of modules that constitute the JUPITER API. Although understanding the structural concept aids in understanding the modules, the modules can be used in applications constructed using different organizational concepts.

The JUPITER API has already been used to construct a number of applications. Three simple example applications are presented herein; for many applications, these simple applications provide a useful starting point. More complicated applications include the universal inverse code UCODE_2005 (Poeter et al., 2005), the multi-model analysis code MMA (Eileen P. Poeter, et al., written commun., 2006; Poeter and Anderson, 2005), and a code named OPR_PPR (Matthew J. Tonkin, SS Papadopulos and Associates, written communication, 2006; Tiedeman and others, 2003, 2004) for assessing data importance to process-model predictions using the local sensitivity Observation-PRediction (OPR) and Parameter-PRediction (PPR) statistics.

Acknowledgments

Substantial guidance and assistance was provided by Justin Babendreier, Karl Castleton, and Steven Fine of the USEPA, and Steve Markstrom, George Leavesley, and Arlen W. Harbaugh of the USGS. Charles Heywood and Claire Tiedeman of the USGS and Laura Foglia, a student at Eidgenössische Technische Hochschule (Swiss Federal Institute of Technology) in Zurich, Switzerland, provided much-appreciated testing of JUPITER applications during development of the API. Matthew Tonkin of Papdopulos and Associates, Inc, under contract with the USGS, used the API to construct the OPR_PPR application and provided valuable comments.

Chapter 2: The JUPITER API Modules and Underlying Concepts
By Edward R. Banta, Mary C. Hill, John E. Doherty, and Eileen P. Poeter

This chapter introduces the JUPITER API modules and describes the organizational structure on which the modules are based. This chapter also discusses the dependence structure among the modules and defines public and private data and subroutines, functions, and generic interfaces.

Data, subroutines, functions, and generic interfaces are organized by grouping them into modules, where each module is designed either to support a particular purpose or to provide a set of general-purpose capabilities. In accordance with Fortran terminology, the term "subprograms," as used in this report, encompasses subroutines and functions.

The modules described in this document are listed in Table 2-1, along with the name of the source-code file and primary purpose for each module. The individual modules are documented in Chapters 4 through 14.

Table 2-1. Modules of the JUPITER API.
[--, none]

Chapter	Appendix	Module (Module Identifier) or Task	Source-code file	Purpose	Authorship of chapter[1]
4	B	Datatypes (TYP)	typ.f90	Definition of derived data types; generic interfaces that nullify pointers in derived data types and deallocate memory associated with derived data types. This module contains no data.	Banta
5	C	Global Data (GLO)	gdt.f90	Data generally needed by many modules.	Banta and Doherty
6	D	Utilities (UTL)	utl.f90	Data, subprograms, and generic interfaces designed to fulfill needs shared by more than one module. Subprograms and interfaces are organized into four categories: Input utilities, output utilities, manipulate utilities, and error-processing utilities. Data used to the manage unit numbers associated with data-exchange files.	Banta, Doherty, and Poeter
7	E	Basic (BAS)	bas f90	Data and subroutines common to many model-analysis applications.	Banta and Doherty
8	F	Model Input-Output (MIO)	mio f90	Data and subroutines needed to communicate data between the application and the process model.	Doherty and Banta
9	G	Equation (EQN)	eqn.f90	Data and subroutines to support the use of equations to define combinations and transformations of parameters, prior information, or model-calculated values.	Doherty

Chapter	Appendix	Module (Module Identifier) or Task	Source-code file	Purpose	Authorship of chapter[1]
10	H	Dependents (DEP)	dep.f90	Data, subprograms, and generic interface for basic processing of model-calculated values.	Banta and Poeter
11	I	Prior-Informa-tion (PRI)	pri.f90	Data and subroutines to support use of prior information in parameter-estimation methods	Banta and Poeter
12	J	Parallel-Processing (PLL)	pll.f90	Data and subprograms to support concurrent execution of model runs to take advantage of a multiprocessor computing environment.	Banta
13	K	Sensitivity (SEN)	sen.f90	Data and subroutines for generating the derivatives of model-calculated values with respect to parameters.	Banta and Poeter
14	L	Statistics (STA)	sta.f90	Data and subroutines to compute commonly needed statistics, especially related to model fit, prior information, and parameters.	Poeter and Hill
15	--	Parameter-Value Generator	--	Application specific. Developer-written modules are discussed that would serve the Generate Parameter Values task. Examples include modules designed to update parameter values using gradient or global-search methods. When the Sensitivity Module assigns sets of parameter values to be used in calculating sensitivities by perturbation, it acts as a parameter-value generator.	Hill, Banta, Poeter, and Doherty

[1] The last names of the authors are listed. The full names and affiliations are Edward R. Banta and Mary C. Hill, U.S. Geological Survey; Eileen P. Poeter, Colorado School of Mines and International Ground Water Modeling Center; and John E. Doherty, Watermark Numerical Computing, Inc., and the University of Queensland, Australia.

Applications, Tasks, and Control

A precept of the JUPITER API is that the API is designed to construct applications (computer programs) involved in the analysis of mathematical and numerical models. The types of model analysis for which the API is designed include, but are not limited to, sensitivity analysis, data needs assessment, calibration, prediction analysis, evaluation of uncertainty, and system management. In this documentation the term "process model" refers to a model designed to simulate a process, typically a physical, chemical, geological, or biological process. The process model may be an external program, or it may be implemented as a subroutine or function of the application. In either case, execution of the process model takes place in a framework designed by the application programmer.

Many of the modules of the JUPITER API are designed with the assumption that the application framework will need to incorporate certain "tasks" to accommodate the interaction between the application and the model, execution of the model, and the model analysis. Applications are envisioned to commonly require one or more of the tasks described in this section within their frameworks, although parts of the API may be useful in applications that do not conform to this set of tasks.

Many model analyses require that selected tasks be executed repeatedly using loops. The loops are called control loops here, and each execution of a control loop is considered to have an associated job. The job of each execution of a control loop is identified by a variable herein called a "control-loop job specifier." A list of control-loop job specifiers can be stored in a character array and used as defined by a computer application. The control-loop job specifiers included in the JUPITER API are listed in Table 2-2. Additional control-loop job specifiers are commonly added for individual computer applications and also could be added to future versions of the JUPITER API.

For example, a program might be organized as listed here and shown in Figure 2-1. The JUPITER API modules provide support for these tasks, as shown in Table 2-1.

- **Initialize:** Read input, allocate memory, and perform preparatory steps as needed.
- Start control loop (Steps of the control loop are numbered as in Figure 2-1; control loops can be parallelized using the capabilities of the Parallel-Processing Module)
    1. **Define Job:** Assign a control-loop job specifier to define the job of the current iteration of the control loop; example job specifiers are listed in Table 2-2. As the control loop executes iterations, jobs in the sequence may be repeated or skipped as required.
    2. **Generate Parameter Values:** Populate an array with one or more sets of parameter values. For example, the JUPITER API provides tools for generating parameter values needed to calculate perturbation sensitivities, for which parameter values are perturbed one at a time for a sequence of executions of the control loop.
    3. **Adapt Parameter Values:** Convert a set of parameter values to a form usable by the process model. When the process model is an external program, this task generally involves creating one or more of the process-model input files using one or more Template files.
    4. **Execute Model:** Invoke the process model, possibly using one of the capabilities provided in the JUPITER API.
    5. **Extract Model-Calculated Values:** Populate an array with selected values read from process-model output files. When the process model is an external program, this task involves reading one or more application-model output files to obtain the values. This is accomplished using one or more Instruction files. When the model is implemented as a subroutine, this task may involve storing values such that they are easily accessible; the JUPITER API provides no special capabilities.
    6. **Use Extracted Values:** This task performs data manipulation using model-calculated values. For example, calculation of perturbation sensitivities from parameter sets and model-calculated values is done in this task.
- End control loop: Loops commonly end here or after the following Evaluate task, or some of the Evaluate task may occur within the control loop while some of the task follows the control loop.
- **Evaluate:** This task performs application-dependent analyses.
- **Cleanup:** Delete unneeded files and (or) deallocate memory.

7

In JUPITER API applications, control loops are application-specific and require programming by the application developer. The programming of control loops can be achieved using computer languages other than the Fortran-90 used to program the existing components of the JUPITER API.

Applications that use the JUPITER API can incorporate all or some of the tasks and loops mentioned above, or additional loops may be defined.

Process models can be designed to facilitate interactions with applications constructed using the JUPITER API. For example, the observation and parameterization capabilities available in MODFLOW-2000 and MODFLOW-2005 simplify construction of Template and Instruction files and allow a JUPITER API application to create process-model input files and read process-model output files more rapidly than would otherwise be possible.

Applications that use the JUPITER API can incorporate all or some of the tasks and loops listed above. Situations for which the order of execution is important are noted in the descriptions of subprograms in Chapters 7 through 14.

Typically, an application that uses the JUPITER API and has a Control Loop would have a character array containing a sequence of control-loop job specifiers. In the SENSITIVITY_EXAMPLE application described in Chapter 18, this array is the CTRLJOB array.

The options for the Control-Loop job specifier defined in this documentation and their corresponding meanings are listed in Table 2-2. Additional, application-specific job specifiers would commonly be defined in applications that use a Control Loop. Support for additional job specifiers also could be added to future versions of the JUPITER API. Of the Control-Loop job specifiers listed in Table 2-2, the SENSITIVITY_EXAMPLE application described in Chapter 18 uses "FORWARD", "FORWARD&SENS", "SENSITIVITY", and "STOP".

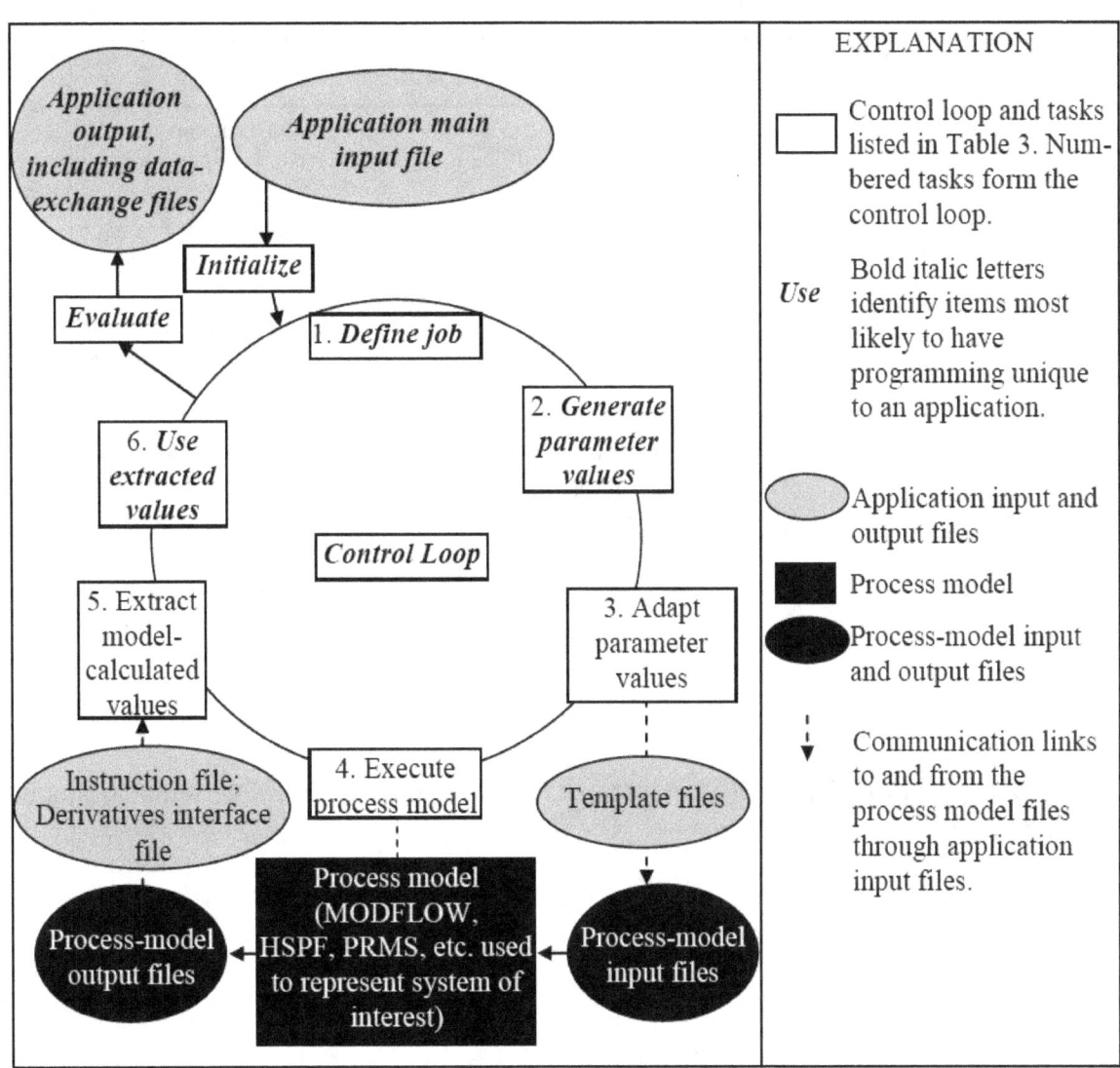

Figure 2-1. Diagram showing how an application (computer program) constructed using the JUPITER API might proceed.

Table 2-2. Control-Loop job specifiers.
[Conventional specifiers and meanings are listed; some specifiers are recognized by modules of the JUPITER API. The full set of Control-Loop job specifiers for a given application would be application-defined]

Control-Loop job specifier	Meaning
"FORWARD"[1]	Run model in forward mode, with current (unperturbed) parameter values. Generally, the purpose of the forward model run is to generate one set of dependent values.
"FORWARD&SENS"[1,2]	Run model with current (unperturbed) parameter values so as to generate dependent values and model-calculated sensitivities for the dependent values with respect to parameters defined in the model. This option requires that the process model be capable of calculating sensitivities and writing them in a form readable according to a protocol specified in a Derivatives Interface file (Appendix A).
"SENSITIVITY"	Run model in forward mode, with current (unperturbed) parameter values and (or) perturbed parameter values as needed to generate sensitivities of dependent values to parameters by perturbation. The Sensitivity Module (Chapter 13) currently supports two finite-difference methods of calculating perturbation sensitivities.
"PRIORSENS"	Populate a sensitivity matrix for prior-information equations (Chapter 11).
"STOP"	Stop execution of the Control Loop.

1. Specifier is recognized by the Basic Module (Chapter 7).
2. Specifier is recognized by the Sensitivity Module (Chapter 13).

Underlying Organizational Structure and Naming Convention for Subroutines, Functions, and Generic Interfaces

Modules contain data, subroutines, functions and(or) generic interfaces related to a common purpose. Table 2-3 illustrates support provided by the JUPITER API modules for the tasks listed previously in the last section and in Figure 2-1.

Table 2-3 also defines the conventions used to name the subroutines, functions and generic interfaces in each module. The conventions are as follows.

1. Each row in Table 2-3 represents a module. Each module has a three-character identifier that forms the first part of each name.
2. Each column in part A of Table 2-3 corresponds to a task and has a three-character identifier that generally is used to form the second part of each name.
3. A final part of the name is often included to clarify the purpose of the subroutine, function, or generic interface, or because the X in Table 2-3 represents multiple components.
4. The parts of the names are separated by an underscore ("_") character.

For example, the X for the Basic Module in the column of Table 2-3 for the Initialize task represents three subroutines that are named BAS_INI_GETARG, BAS_INI_GETOPTIONS, and BAS_INI_MODELEXEC. In the column for the Generate Parameter Values task, the X represents a single subroutine, which is named BAS_GEN. Some modules listed in the upper part of the table include components designed to be used in multiple tasks; the names of these subprograms do not include a task identifier.

Modules that serve purposes not directly related to specific tasks are listed in part B of Table 2-3. In these modules, the subprogram name is composed of the three-character module identifier followed by an underscore and additional characters.

Table 2-1 shows that the capabilities provided for the "Generate Parameter Values" task are minimal, and that task is omitted from Table 2-3. The JUPITER API does not include, for example, local regression methods or global search methods for generating new sets of parameter values. The JUPITER API does not contain such capabilities because there are no standard methods that are likely to be used in a large number of JUPITER API applications. For example, the JUPITER API application UCODE_2005 contains three regression methods, all of which were modified substantially from previous versions during development of UCODE_2005. The JUPITER API is limited to less dynamic aspects of the programming needed for such applications.

Users of the JUPITER API can take advantage of capabilities developed for earlier open-source, public-domain applications. For example, the nonlinear regression modules and the control-loop programming in UCODE_2005 might be of considerable use to developers of applications that use other parameter estimation algorithms, applications that optimize management scenarios, or other applications that require adjustment of parameters and evaluation of calculated dependents.

Table 2-3. Tasks supported by JUPITER API modules.
[Each X indicates that the corresponding module includes support for the corresponding task. Identifiers of tasks normally within the Control Loop are white on a gray background. The numbers on the task descriptions correspond to the task numbers in the text and in Figure 2-1.]

A. Modules with data, subroutines, functions, and generic interfaces that each primarily serve one task[1]										
Module	**Module identifier[1]**	**Task and Task Identifier[1]**								
		Initialize	1. Define job	2. Generate parameter values	3. Adapt parameter values	4. Execute model	5. Extract model-calculated values	6. Use extracted values	Evaluate	Clean-up
		INI	DEF	GEN	ADA	EXE	EXT	UEV	EVA	CLN
Basic	BAS	X		X		X				X
Model Input-Output	MIO	X			X		X			X
Equation	EQN	X								X
Dependents	DEP	X					X	X		X
Prior-Information	PRI	X	X					X		X
Parallel-Processing	PLL	X			X	X	X			X
Sensitivity	SEN	X	X	X		X		X		X
Statistics	STA	X						X	X	X
B. Modules with more generally applicable data and subprograms[2]										
Module[2]	**Module Identifier[2]**	**General description of module data and (or) subprograms**								
Datatypes	TYP	Subroutines for nullifying pointers in derived data types and for deallocating memory associated with derived data types. This module contains no data.								
Global Data	GDT	Data generally needed by many modules. This module contains no subprograms.								
Utilities	UTL	Subprograms and data designed to fulfill needs shared by more than one module. Subprograms are organized into five categories: Data-exchange file management, input utilities, output utilities, manipulate utilities, and error-processing utilities. The data contained in this module relates to managing unit numbers of data-exchange files (Chapter 17).								

[1] Names of subroutines, functions, and generic interfaces in the modules are constructed as the module identifier, the task identifier, and, optionally, additional characters. The parts of the name are separated by an underscore character (for example, BAS_INI_GETARG, BAS_INI_GETOPTIONS, and BAS_GEN).

[2] Names of subroutines, functions, and generic interfaces in the modules are constructed as the module identifier, an underscore character, and additional characters (for example, UTL_GETLINE and UTL_READ2D).

Module Dependence

Some JUPITER API modules can be used independently. Other modules require data or procedures from other JUPITER API modules and are said to depend on those modules. Use of these modules requires that the modules upon which they depend be included when compiling the application. Table 2-4 shows the dependence relations among modules of the API, and thus shows which other modules are needed when a particular module is used in an application.

In Fortran 90 a module or a subprogram accesses data or subprograms in other modules through a Fortran 90 USE statement. Such a reference establishes what is called USE association. A rule of USE association prevents a module from referencing itself, either directly or indirectly. For example, if a subprogram in module A uses module B, and a subprogram in module B uses module C, neither module B nor module C can use module A. As a result of this restriction, there is a hierarchy of dependence among modules that use other modules. Table 2-4 shows the dependence relations among modules of the JUPITER API. The purpose of Table 2-4 is to help programmers determine which other modules are required when a particular module is to be used in an application. Modules in lower rows are dependent on modules in higher rows. Modules in a row can USE modules in preceding (higher) rows. Modules do not USE other modules in the same or later (lower) rows. The syntax for the USE statement to use each module is listed in Table 2-5.

Table 2-4. Module dependence.

Modules independent of other modules	**Datatypes Module**	**Global Data Module**	
Modules that depend on other modules. **Modules in lower rows can depend on modules in the rows above.**	**Utilities Module** uses Datatypes and Global Data Modules		
	Basic Module uses Datatypes, Global Data, and Utilities Modules	**Model Input-Output Module** uses Datatypes, Global Data, and Utilities Modules	**Equation Module** uses Global Data and Utilities Modules
	Dependents Module uses Datatypes, Global Data, Utilities, and Equation Modules	**Prior Information Module** uses Datatypes, Global Data, Utilities, and Equation Modules	**Parallel-Processing Module** uses Datatypes, Global Data, Utilities, and Model Input-Output Modules
	Sensitivity Module uses Datatypes, Global Data, Utilities, Basic, and Dependents Modules	**Statistics Module** uses Datatypes, Global Data, Utilities, and Dependents Modules	

Table 2-5. Module names for USE statements.
[The module name for USE statements in some cases differs from the name used in chapter and appendix titles]

Module	Name for USE statements	Module	Name for USE statements
Datatypes	DATATYPES	Dependents	DEPENDENTS
Global Data	GLOBAL_DATA	Prior Information	PRIOR_INFORMATION
Utilities	UTILITIES	Parallel-Processing	PARALLEL_PROCESSING
Basic	BASIC	Sensitivity	SENSITIVITY
Model Input-Output	MODEL_IO	Statistics	STATISTICS
Equation	EQUATION		

Public and Private Data and Subprograms

Within each module, data needed within the module but generally not by other modules are protected from corruption by other program units by making them private using the PRIVATE statement. Similarly, subprograms not expected to be needed by other modules are declared as PRIVATE. Data and subprograms that may be used or invoked from outside a module are declared in a PUBLIC statement. As is required by Fortran 90, access to data or subprograms declared as PUBLIC in modules is provided by naming the module in a USE statement, as discussed above. The public data and subprograms and the private data of the API are described in the appendix corresponding to each module (Table 2-4). The private subprograms are not described in this documentation because they are not expected to be directly invoked by applications built with the JUPITER API.

The public data and subroutines are documented by listing their names and purposes in tables in the chapters for each module (Chapters 4 through 14) and by providing additional information in appendixes. Private data are described in the Appendices. Each appendix (Appendix B through L) corresponds to one of the 11 chapters, as shown in Table 2-1.

Chapter 3: Data Input and Output Mechanisms
By Edward R. Banta, John E. Doherty, Mary C. Hill, and Eileen P. Poeter

This chapter describes the capabilities provided by the JUPITER API for input of data and output of analysis results and data. Data input and output mechanisms provided by the API are listed in Table 3-1.

Data Input Mechanisms

Input blocks are the basic input structure of the JUPITER API. Applications commonly are designed such that input may be contained in a single main input file. However, input blocks allow application developers and end users to structure input flexibly. Within selected input blocks, names of three other types of input files can be specified: a Derivatives Interface File, Template Files, and Instruction Files. Additional input files can be defined by the application developer.

This section describes input blocks in detail, because they are referenced in many of the following chapters. Then, this section briefly describes derivative interface files and template and instruction files; details of these files are presented in Appendix A.

In explanations of input requirements, an input item shown in ALL CAPITALS is a keyword that must be entered literally; lower case and combined upper and lower case are used to represent a user-defined value or a character string that may be one of a set of alternative keywords. Subroutine arguments also are shown in ALL CAPITALS; the context will distinguish keywords from subroutine arguments. Input items in square brackets are optional.

Input Blocks

Input blocks, as defined by the JUPITER API, are designed to accommodate the input requirements of individual applications, while maintaining flexibility to facilitate input-file preparation by users, who may be using multiple JUPITER-based applications to analyze a given modeling problem. Developers will find that coding an application to read its input in the form of input blocks is straightforward and convenient, whether the input is to be used to populate scalar variables or arrays. Users will find that the consistent syntax used in input blocks aids in the construction and comprehension of input files.

The use of input blocks requires little programming on the part of the application developer. The Utilities Module (Chapter 6 and Appendix D) provides one subroutine for reading an input block (UTL_READBLOCK) and two subroutines for accessing data read from an input block (UTL_FILTER and UTL_FILTERLIST). The basic functionality of input blocks is implemented by these three subroutines. The Utilities Module provides a small number of other subroutines for manipulating data read from input blocks. See the "Example Application: SENSITIVITY_EXAMPLE" section of Chapter 18 for examples of input blocks and the use of these subroutines.

Chapter 3: Data Input and Output Mechanisms

Table 3-1. Mechanisms provided by the JUPITER API for data input and output.

Mechanism	Purpose	Module(s) that use the mechanism
Input		
Input block	Input of any or all data needed by a JUPITER API application.	All modules that read data. Once the data are read, any module has access to the data.
Derivatives Interface file	Defines how to read sensitivities from process-model output files.	Basic and Sensitivity Modules.
Template files	Construction of model-input files using changed values of parameters.	Model Input-Output Module.
Instruction files	Defines how to read simulated values from process-model output files.	Model Input-Output Module.
Output		
Output file(s)	Designed primarily to be read by user.	Many. Many modules can contribute to the same output file.
Output and Input		
Data-exchange files	Designed primarily to be read by other applications, software for creating graphics, and so on. Usually consist of a line of column labels followed by columns of data.	Many.
Matrices	Correlation of weights and application-dependent uses. Matrices can be compressed. The API includes utilities for manipulating matrix data stored in compressed or uncompressed form in memory or in files.	Produced and read by the Basic and Utilities Modules. Once read, the matrices may be used by any module.

Data in input blocks are identified using keywords. Some keywords are defined by the JUPITER API; others are defined by the application programmer. Though keywords of the API are not case sensitive, in this report CAPITAL LETTERS are used to identify keywords that need to be entered literally. Input items that include lower-case letters need to be replaced by the user. Input items listed in square brackets are optional.

Using these conventions, input blocks have the basic structure:

```
BEGIN Blocklabel [Blockformat]
Blockbody
END Blocklabel
```

where,
BEGIN is a JUPITER API keyword that defines the first line of an input block.

Blocklabel identifies the input block and defines the keywords that are recognized when listed in the blockbody. Some Blocklabels are defined by the JUPITER API; others are added by the application programmer. Blocklabels defined by the JUPITER API include, for example, Observation_Data and Parallel_Control. An application-specific Blocklabel for a Monte Carlo application might be, for example, Parameter_Distribution.

Blockformat is optional and defines the format used to read the input block. The options for blockformat are KEYWORDS, TABLE, and FILES, as described in the section "Blockformats". If omitted, the default is KEYWORDS. The blockformat options provided by the JUPITER API are very flexible, and it is not expected that application-specific blockformat options would be defined.

Blockbody refers to one or more lines of text that contain either data or names (or pathnames) of files from which data are to be read. Content and format of the blockbody depend on the specified blocklabel and blockformat and, possibly, other specifications and options. Data read can be used to populate scalars or arrays of type integer, real, double precision, logical, or character.

END is a JUPITER API keyword that indicates the end of the input block. The blocklabel following the **END** keyword needs to match the blocklabel specified in the corresponding **BEGIN** line.

As described below, comment lines, indicated by the presence of a "#" character in the first column of the line, may be used in some places in input blocks. All non-comment lines may be indented as desired to aid readability.

Two Forms of Input Blocks

The JUPITER API provides tools that support two forms by which an end user may organize data within an input block. One form, selected by choosing (or defaulting to) blockformat KEYWORDS, generally is most advantageous when the data are used to populate scalar variables; the second form, selected by choosing blockformat TABLE, generally is most advantageous when the data are used to populate arrays.

In the first form, data are organized in a single list of keywords and their associated values using Blockformat KEYWORDS. For example, a complete input block might be:

```
BEGIN MY_APPLICATION_OPTIONS KEYWORDS
  INPUTFILE = C:\MyApplication\inputfile.txt
  WARNLEVEL = 4    DOANALYSIS = true
END MY_APPLICATION_OPTIONS
```

The Blocklabel "MY_APPLICATION_OPTIONS" and the three keywords "INPUTFILE", "WARNLEVEL", and "DOANALYSIS" are defined by the application involved, using subroutines provided as part of the JUPITER API. The order of the keyword=value entries is not important unless a keyword is repeated. When a keyword is repeated, the utility that retrieves data from an input-block data structure returns the value associated with the final occurrence of the keyword. Lines may contain any number of entries; in the example above, the first line of the blockbody contains one keyword, and the second line contains two keywords.

In the second form, data are organized such that selected attributes are associated with each occurrence of a special type of keyword referred to as a keyitem. For example, an application may read many parameters and require a list of attributes for each parameter. The keyitem in this situation would define the parameter name, and attributes would be associated each parameter name. Commonly, the blockformat option TABLE is used in this circumstance. For example, a complete input block might be:

```
BEGIN PARAMETER_DATA TABLE
# Comment: Input block to define parameter attributes
  NROW=3 NCOL=4 COLUMNLABELS
  PARAMNAME    STARTVALUE   ADJUSTABLE  UNITS
  par1         10.3         true        --
  par2         1.7          true        meters
  par3         9.2          false       --
END PARAMETER_DATA
```

This example uses the JUPITER API "PARAMETER_DATA" Blocklabel. NROW and NCOL are JUPITER API keywords that define the dimensions of the table. COLUMNLABELS is a JUPITER API keyword needed if keywords are to be placed at the top of each column of the table. PARAMNAME, STARTVALUE, and ADJUSTABLE are JUPITER API keywords; UNITS is an application-defined keyword. The API provides a generic interface that the application can invoke to populate four arrays with the contents of the four columns. The length of each array equals the number of rows, which equals the number of times the keyitem is defined. Here, the input block defines three parameters and three associated attributes for each parameter.

Groups

The JUPITER API provides support for populating arrays of default attributes for keyitems. This is achieved by defining groups an assigning members to groups. Using the PARAMNAME example above, parameter par1 might be a member of one group and par2 and par3 might be members of another group. The groups can be used to define default attribute values for group members, which can simplify data input.

To define groups, typically a pair of input blocks is used: one is a "group" input block and the other is a "members" input block. The group input block assigns default attributes to groups. The members input block defines the members of each group and overrides the defaults as needed. Use of groups can simplify preparation of input considerably.

The keyword "GROUPNAME" has special meaning in defining groups. In the group input block, GROUPNAME must be the keyitem. In the members input block, GROUPNAME is not the keyitem; it is an attribute. The UTL_GROUPLIST subroutine (Appendix D) is used to apply the default attributes defined in the groups input block to members of the group as defined in the members input block. This capability has the potential to simplify preparation of input considerably.

The following example illustrates the use of groups:

```
BEGIN Observation_Groups KEYWORDS
  GROUPNAME=HeadObs
    STATFLAG=var
    STATISTIC=1.5
  GROUPNAME=FlowObs
    STATFLAG=cv
    STATISTIC=0.08
END Observation_Groups

BEGIN Observation_Data KEYWORDS
  OBSNAME=Head1  GROUPNAME=HeadObs  OBSVALUE=101.8
  OBSNAME=Head2  GROUPNAME=HeadObs  OBSVALUE=98.6    STATISTIC=1.0
  OBSNAME=Head3  GROUPNAME=HeadObs  OBSVALUE=149.3
  OBSNAME=Flow1  GROUPNAME=FlowObs  OBSVALUE=120.0
  OBSNAME=Flow2  GROUPNAME=FlowObs  OBSVALUE=65.0
END Observation_Data
```

In this example, the Observation_Groups input block defines two groups, named HeadObs and FlowObs. Values for STATFLAG and STATISTIC are defined for each group. The Observation_Data input block defines two observations as members in one group, and three in the other. An application designed to read these two input blocks would use UTL_GROUPLIST to use the values from the Observation_Groups input block as defaults for group members define in the Observation_Data input block. After execution of UTL_GROUPLIST, attributes would be associated with group members as follows:

```
OBSNAME    GROUPNAME    OBSVALUE      STATFLAG    STATISTIC
-------    ---------    ------------  --------    ---------
Head1      HeadObs      101.80        var         1.5
Head2      HeadObs      98.600        var         1.0
Head3      HeadObs      149.30        var         1.5
Flow1      FlowObs      120.00        cv          0.80E-01
Flow2      FlowObs      65.000        cv          0.80E-01
```

The "Example Application: GROUP_EXAMPLE" section of Chapter 18 describes a simple example program that reads the two input blocks and generates the tabular output above.

Blockformat

The blockformat of an input block defines the format used for organizing the blockbody. This section describes the blockformat options in detail. The options and corresponding formats are listed in Table 3-2.

Table 3-2. Blockformat options for input blocks defined by the JUPITER API.

Blockformat	Prescribed input format
KEYWORDS	**Blockbody** entries are of the form **keyword**=value and need to be separated by one or more delimiters or a new line; delimiters include spaces, commas, and tab characters. Comment lines[a] are allowed anywhere in the blockbody. Character strings that are not in the form **keyword=value** are ignored.
TABLE[b]	**Blockbody** consists of a table of data in rows and columns, which may have labels on the columns and may be read from the main input file or from another input file. Comment lines[a] are allowed right after the BEGIN statement but not in the rest of the blockbody. Options can be defined using arguments on the first line of the input block. Options supported by the JUPITER API include whether to use column labels or a default column order, or to read tables of data from other files.
FILES[b]	**Blockbody** consists of a list of one or more file or path names such that one input block is read from each file. Comment lines[a] between file names are allowed. Each of the referenced files needs to contain an input block with the same **Blocklabel** as the input block from which it is referenced. The input block in each file can be read using any of the three **blockformat** options.

[a] Comment lines are lines starting with a "#" character as the first character in the line. Comment lines can be located only as indicated, and are ignored.
[b] The TABLES and FILES options can be used to read data from separate files.

Blockformat KEYWORDS

If **blockformat** is not specified or if **blockformat** is specified as **KEYWORDS**, then **blockbody** is expected to be a series of entries of the form **keyword=value**. **Value** may be a character string, an integer, or a floating-point number. When the keyword requires a logical variable, **value** may be "true," "yes," "t," or "y" (all of which are synonymous) or "false," "no," "f," or "n" (which are synonymous). For example, the blockbody will be expected to contain such entries as "OBSNAME=head1", "ESTIMATE=true", "PLOTSYMBOL=1", or "STATISTIC=0.5". There can be spaces on each side of the equal sign, but the entry must not be split across lines. New lines can be used to separate entries; alternatively, consecutive entries, separated by one or more delimiters, can be placed on the same line. In general, spaces, commas, and tab characters are interpreted as delimiters separating **keyword=value** entries. Single ($'$) or double ($''$) quotes may be used to enclose either the entire **keyword=value** entry or just **value**. When quotes are used, **value** may contain embedded spaces and tab characters. Spaces and (or) tab characters may be inserted between the **keyword** and the "=" sign or between the "=" sign and the **value** for aesthetic purposes, but neither spaces nor tab characters are required.

Blockformat TABLE

When **blockformat** is specified as TABLE, the data to be read are expected to be organized in rows and columns. Depending on the arguments NDEFCOLS and COLNAMES in the call to UTL_READBLOCK, a default column order may be defined. To specify a default column order, NDEFCOLS would be specified as the number of default columns and COLNAMES would contain the column names. If NDEFCOLS is specified as 0, the input block does not have a default column order.

If blockformat is specified as TABLE, the first line of **blockbody** is of the form:

  NROW=nr **NCOL**=nc [**COLUMNLABELS**] [**DATAFILES**=nfiles] [**GROUPNAME**=gpname]

where:

NROW and **NCOL** are required keywords.

nr is the number of rows of data values.

nc is the number of columns of data values.

COLUMNLABELS is an optional keyword, but it must be present if the input block does not have a default column order defined.

    If **COLUMNLABELS** is omitted: The default column order defined for the input block is used to identify the columns of data. The data columns need to be in the default column order.

    If **COLUMNLABELS** is present: Keywords are used as column labels. The second line of the blockbody must consist of **nc** keywords. The columns can be ordered as desired as long as the data in each column is consistent with the keyword used as the columnlabel.

DATAFILES is an optional keyword.

    If **DATAFILES** is omitted: **nr** rows of data are read from the input block as shown in the left-hand column of Table 3-3. The data type expected for **val** depends on the **blocklabel** and possibly on **column-name**. All data values for a row need to be on one line of the file.

    If **DATAFILES** is present: A list of file names is read, as shown in the right-hand column of Table 3-3. **Nfiles** is an integer indicating the number of file names in the list. Each **filename** is the pathname or file name of a file from which rows of data are to be read. File names with spaces need to be enclosed in double quotes. Each file must contain zero or more rows of data in columns in either the default column order or the order defined by the **column-name** entries, if **COLUMNLABELS** is specified. Data read from all files are combined as if read from one file. Each file is read in order until **nr** rows of data have been read. If **SKIP**=nskip is specified, **nskip** lines at the beginning of the file are ignored, and reading of data starts on the following line.

GROUPNAME is an optional keyword.

    For "members" blocks that support groups (see the "Groups" section, above), **GROUPNAME**=gpname can be used to assign a group name to all rows included in the input block; **gpname** is the group name. If **GROUPNAME**=gpname is present: (1) If **COLUMNLABELS** is omitted, **GROUPNAME** must not be in the default list of columns; and (2) If **COLUMNLABELS** is present, **GROUPNAME** must not be listed as a column name.

Table 3-3. Use of DATAFILES option of blockformat TABLE.
[When blockformat is specified as TABLE, the format of blockbody after the first line without and with the optional keyword DATAFILES is shown in the left and right column, respectively. column-name, a keyword that identifies the data in the associated column; val, data value; nskip, number of lines to skip at the top of the file.]

Without **DATAFILES** keyword	With **DATAFILES** keyword

`[column-name] [column-name] ...`
`    val          val       ...`
`    val          val       ...`
`    ...          ...       ...`
	`[column-name] [column-name] ...`
	`filename  [ SKIP=nskip ]`
	`filename  [ SKIP=nskip ]`
	`    ...`

`number of lines: `**`nr`**	`number of lines: `**`nfiles`**

Two equivalent examples of input blocks using blockformat TABLE without the DATAFILES keyword follow:

```
BEGIN Model_Input_Files TABLE
# Note that default column order
# is used: MODINFILE TEMPLATEFILE
  NROW=3  NCOL=2
  .\data\tc1fwd.sen    tc1fwdsen.tpl
  .\data\tc1sen.sen    tc1sensen.tpl
  .\data\tc1gwt.txt    tc1gwt.tpl
END Model_Input_Files
```

```
BEGIN Model_Input_Files TABLE
  NROW=3  NCOL=2  COLUMNLABELS
  MODINFILE            TEMPLATEFILE
  .\data\tc1fwd.sen    tc1fwdsen.tpl
  .\data\tc1sen.sen    tc1sensen.tpl
  .\data\tc1gwt.txt    tc1gwt.tpl
END Model_Input_Files
```

In the following example input block, the DATAFILES keyword is used to indicate that input data are to be read from two files named in the input block:

```
BEGIN observation_Data TABLE
  NROW=32  NCOL=5  DATAFILES=2  COLUMNLABELS
  obsname  obsvalue    statistic  statflag   GROUPNAME
  tc1a.hed
  tc1b.hed
END observation_Data TABLE
```

Each of the referenced files contains only rows and columns of data. For example, the first 17 rows of data are contained in the file tc1a.hed (see below) while the remainder will be read from tc1b.hed.

```
h1.0     101.8040        1.0025      var     Heads1
h1.1    -0.2900000E-01   0.0025      var     Heads1
h1.12   -0.1290000       0.0025      var     Heads1
h2.0     128.1170        1.0025      var     Heads1
h2.1    -0.4100000E-01   0.0025      var     Heads1
h2.2    -0.5570000       0.0025      var     Heads1
h2.8    -11.53100        0.0025      var     Heads1
h2.12   -14.18400        0.0025      var     Heads1
h3.0     156.6780        1.0025      var     Heads3
h3.1    -4.381000        0.0025      var     Heads3
h3.12   -42.54000        0.0025      var     Heads3
h4.0     124.8930        1.0025      var     Heads1
h4.1    -0.6700000E-01   0.0025      var     Heads1
h4.12   -14.30400        0.0025      var     Heads1
h5.0     140.9610        1.0025      var     Heads1
h5.1    -0.6000000E-01   0.0025      var     Heads1
h5.12   -21.67600        0.0025      var     Heads1
```

Blockformat FILES

If **blockformat** is specified as FILES, the blockbody consists of one or more lines, each containing the name of a file, possibly including a path. The input block structure, from the **BEGIN** line through the **END** line, is recursive; each file listed needs to contain one complete input block. The **blocklabel** of each file needs to be the same as the **blocklabel** for the originating input block. Generally, the **blockformat** specified in the referenced files is not "FILES". Either "KEYWORD" or "TABLES" can be used, and different **blockformats** can be used in different files. This option allows data stored in different **blockformats** to be read by as part of one input block.

Lines with # as the first character are interpreted as comment lines and are ignored.

In the following example, files tc1.hed and tc1.flo each need to contain one complete input block.

```
BEGIN Observation_Data FILES
# heads
  tc1.hed
# flows
  tc1.flo
END Observation_Data
```

Using **blockformat** FILES is the second alternative by which the main input file can point to other input files. The two alternatives are compared in Table 3-4.

Table 3-4. Alternatives for reading data from files identified in the main input file.

	Alternative	
	Blockformat TABLE With DATAFILES	**Blockformat FILES**
Number of input blocks	One. The BEGIN and END lines are located in the main input file.	More than one. BEGIN and END lines are located in the main input file, in each of the files listed in the main input file, and possibly in files listed in those files. Blocklabel needs to be the same for all of these input blocks.
Format of input blocks	All data are read as a table.	Blockformat in the main input file is FILES. In the files, blockformat is usually TABLE or KEYWORDS.

Derivatives Interface File

When a model (for example, MODFLOW-2000, Hill and others, 2000) is programmed to calculate sensitivities of model-calculated values with respect to parameters, the JUPITER API provides a method to take advantage of that capability. The Options input block of the Basic Module (Chapter 7) can be used to read the name of a "derivatives interface file". The Sensitivity module (Chapter 13) uses information read from the derivative interface file to read and store the sensitivities.

Template Files

When a JUPITER application needs to run a process model, it typically needs to modify process-model input files so that they reflect the parameter values defined in the application. User-defined template files are used by a JUPITER application to create the needed model-input filess. As part of the Adapt Parameter Values task (Figure 2-1 and Table 2-3), the JUPITER API provides the ability to use template files to prepare process-model input files. The use and construction of template files is described in Chapter 8 and Appendix A.

Instruction Files

After a model run is made, model-calculated values from model output can be read. These model-calculated values are referred to as dependents and can include simulated equivalents to observations and (or) predictions. User-defined instruction files provide the information needed by the JUPITER API Extract Model-Calculated Values task to extract (read) the model-simulated values from the process-model output files (Figure 2-1 and Table 2-3). The use and construction of instruction files is described in Chapter 8 and Appendix A. Model-calculated sensitivities are not treated as dependents.

Chapter 3: Data Input and Output Mechanisms

Main Output File

For the purposes of this documentation, it is assumed that an application developed using the JUPITER API writes a main output file designed to convey information generated by the application, which could include a summary of the analysis performed by the application, or a highly detailed account of the application's workings, or something in between. Certain subroutines of the JUPITER API are designed to write output to such a file, and many of these allow the application developer or the user to control the amount of output generated. Access to the main output file generally would be defined at the beginning of an application by opening the file for formatted output. Output can be accumulated as different capabilities of different modules are invoked. Subroutines designed to write to the main output file generally have the dummy variable IOUT in the argument list, where IOUT is the unit number on which the main output file has been opened.

Data Input and Output Mechanisms

Data-exchange files and matrix-data files generally are produced by one program with the intent that they be used by another. These are defined briefly here and further in Appendix A.

Data-Exchange Files

Data-exchange files provide a means for communicating data from a JUPITER application to another application, which could be another JUPITER application, a spreadsheet program, or another data-analysis or plotting program. JUPITER data-exchange files contain data with little or no explanatory text because they are intended to be read by another computer program. Data-exchange files are produced by several of the modules described in this document. In the version of the API documented in this report, all data-exchange files are in ASCII (text only) format.

Each type of data-exchange file has a specific format, which in most cases consists of a line defining column headings followed by rows of data formatted in space-delimited columns. In this type of format, each heading consists of one or more words enclosed in double quotes. Other data-exchange files may be structured with each line containing an identifier followed by a numeric or character value. The types of data-exchange files supported by the JUPITER API and the conventions used for their construction are described in Chapter 17. The list presented in Chapter 17 can be considered to be a basic set to which application-specific or API additions are likely.

The data-exchange files are produced with a minimal amount of metadata. If additional metadata are needed, the application developer can include them in one or more additional files.

Matrix Data Files

Many applications require matrices that can not be read conveniently using the methods previously discussed. For example, variance-covariance matrices and weight matrices can not easily be read by the preceding methods. Often these matrices are needed by a sequence of applications, and the matrices may be large but sparse (many elements equal zero).

The JUPITER API supports input and output of matrix data using either of two formats. A full matrix with a value specified for each matrix element can be read, or just the non-zero elements of a matrix can be stored in a file in compressed form and read by the JUPITER API into a storage structure designed to hold a compressed matrix. Matrices can also be written in compressed format. Compressed matrices can be used to reduce the disk space and memory required for storage of large, sparsely populated matrices. The API includes utilities for manipulating matrix data stored in compressed or uncompressed form. See Chapters 6 and 7 and Appendix A for details regarding the use of matrices.

Chapter 4: Datatypes Module
By Edward R. Banta

In Fortran 90, derived data types provide an organizational structure for data. Careful definition and use of derived data types can facilitate a structured approach to programming, if the derived types are used with discretion to encapsulate information into coherent units. In the JUPITER API, a limited number of derived data types are defined to address specific programming issues and to encapsulate data that can be made available to multiple modules and subprograms.

The Datatypes Module defines a small number of widely-used derived-data types. The module also contains subprograms that initialize or deallocate components of structures of derived types, where the components have the pointer and dimension attributes.

Derived data types defined in the Datatypes Module are described in the rest of this section. The derived data types relate to input blocks (Chapter 3), matrices (including two types for the compressed storage described in the "Matrix Data Types" section, below), or arrays of character strings. The data types related to matrices and character strings are used to maximize efficiency of memory use, especially when multiple matrices or multiple arrays of character strings of varied dimensions are required. Specialized data types for matrices support compressed storage of double precision two-dimensional arrays.

The Datatypes Module is accessed by the statement "USE DATATYPES". Subprograms of the Datatypes Module are documented in Appendix B. The source code for the Datatypes Module is in a file named typ.f90.

Input-Block Data Types

Design of the Input-Block Data Types

Input read from input blocks (see "Data Input" section of Chapter 3) initially is stored in structures of the derived data types LNODE and LLIST, which are designed specifically for this purpose. The LNODE and LLIST data types are defined as follows:

```
TYPE :: LNODE
  CHARACTER(LEN=40)                           :: KEYWORD
  CHARACTER(LEN=40)                           :: VALUE
  INTEGER                                     :: NCHAR
  CHARACTER(LEN=1), ALLOCATABLE, DIMENSION(:) :: STRING
  TYPE (LNODE), POINTER                       :: NEXTNODE
END TYPE LNODE

TYPE :: LLIST
  CHARACTER(LEN=12)       :: GROUP
  TYPE (LNODE), POINTER :: LHEAD, LTAIL
  TYPE (LLIST), POINTER :: NEXTLIST
END TYPE LLIST
```

When an unkown amount of data is to be stored, a common approach is to store the data in a "linked list." A linked list typically consists of a series of structures of a derived data type, where each structure (or "node") of the derived type includes one or more components to hold data and a pointer to the next structure of the same derived type. As memory is required to hold new data, additional nodes are allocated incrementally. To read data from an input block and store it, the

27

JUPITER API uses a variation of this approach that can be thought of as a "branched linked list." In this documentation, the term "input-block data structure" commonly is used to refer to this type of data structure. In this approach, a linked list is constructed using nodes of type LLIST. Each node of type LLIST contains a pointer to the next node of type LLIST. Each LLIST node also contains a pointer to a structure of type LNODE. The LNODE data type includes a pointer that can point to another LNODE structure, so the LNODE structures also can be used to form linked lists. Each linked list of LNODE structures can be visualized as a branch connected to a node in a linked list of LLIST structures. A single LLIST structure declared with the POINTER attribute provides access to all contents of the LLIST structures in the main list as well as all the LNODE structures in the branch lists. Each LLIST structure also includes a pointer that is used to point to the last LNODE node (tail) of the branch list; this pointer facilitates addition of new nodes to the end of the branch list without iterating through every LNODE structure in the list. The actual data are stored in the VALUE and STRING components of each LNODE structure. The identifier for each element of data is stored in the KEYWORD component. The STRING component is allocated to the minimum number of elements required to contain the string to be stored. The elements of the STRING component array are populated with characters of that string. VALUE contains the first 40 characters of that string. When a series of data items supports the "group" concept, the group name is stored in the GROUP component of each LLIST structure. A branched linked list that includes four linked lists is illustrated in Figure 4-1, showing the use of pointers. The first list contains two nodes of type LNODE, the second contains four nodes, and the third and fourth contain three nodes each.

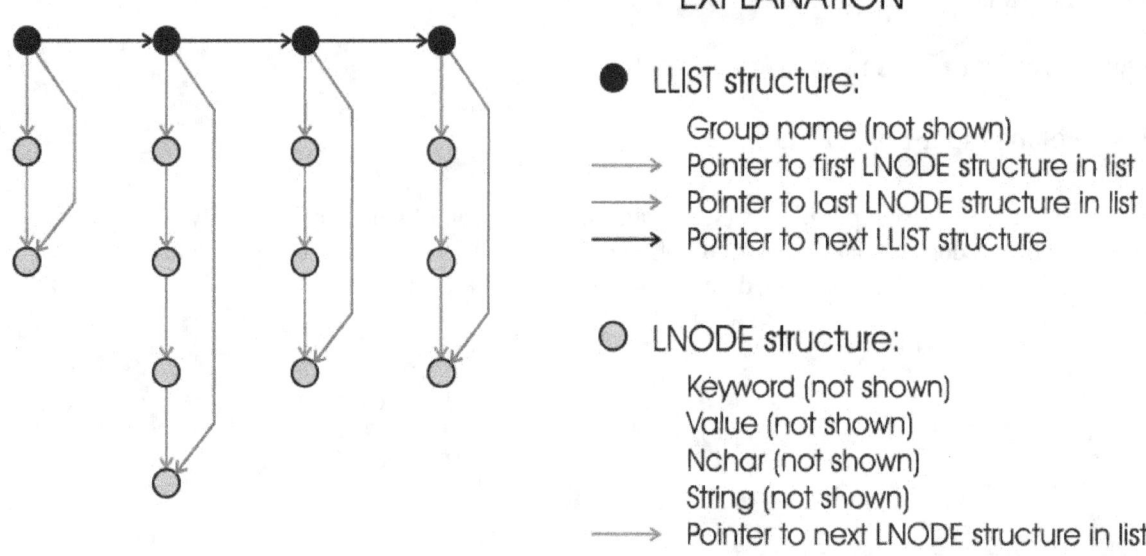

EXPLANATION

● LLIST structure:
  Group name (not shown)
  → Pointer to first LNODE structure in list
  → Pointer to last LNODE structure in list
  → Pointer to next LLIST structure

○ LNODE structure:
  Keyword (not shown)
  Value (not shown)
  Nchar (not shown)
  String (not shown)
  → Pointer to next LNODE structure in list

Figure 4-1. An example of the use of the input-block data types LLIST and LNODE.

Chapter 4: Datatypes Module

Reading and Storage of Data in Input Blocks

Each call to UTL_READBLOCK (Chapter 6 and Appendix D) reads an input block from an input file, unless the REQUIRED argument of UTL_READBLOCK is false and an input block having the specified blocklabel (Chapter 3) is not the next item in the input file open on unit INUNIT. The data read into input-block data structures by UTL_READBLOCK and possibly manipulated by other utility subroutines, can be accessed by calls to UTL_FILTER (to assign a value to a scalar variable), by calls to UTL_FILTERLIST (to populate an array), or by customized subroutines that are designed to traverse an input-block data structure and extract, manipulate, and store data. Each input-block data structure can be considered to belong to the module where the corresponding LLIST pointer is declared. Generally, the variables or arrays to be populated are in the same module where the LLIST pointer is declared. However, in modules of the JUPITER API, the LLIST pointer for each input-block data structure is declared with the PUBLIC attribute. Any application or module with a USE statement that refers to a module in which a PUBLIC LLIST pointer is declared can access the data stored in the associated input-block data structure. As a result, once an LLIST pointer is associated with an input-block data structure and that data structure is populated by a call to UTL_READBLOCK, any subprogram in any module can access the data in that data structure.

Example: Reading an Input Block

This section illustrates two of the utilities designed to read and retrieve data in input blocks. The specifications for constructing input blocks are presented in Chapter 3. The utilities (UTL_READBLOCK and UTL_FILTERLIST) are documented in Chapter 6 and Appendix D. Assume that a file named "block.in" contains the following input block:

```
BEGIN STATE_DATA KEYWORDS
  STATENAME=Colorado
    CAPITAL=Denver
    ORIGINAL13=NO
    POPULATION=4301261
    STATEHOODYEAR = 1876
  STATENAME = "New Jersey"
    CAPITAL=Trenton
    STATEHOODYEAR = 1787
    ORIGINAL13 = YES
    OCEAN = Atlantic
    POPULATION = 8414350
END STATE_DATA
```

The following program reads the input block, stores the data in an input block structure associated with pointer BLOCKPTR, populates arrays of various types, and writes the data to an output file. To compile this program, the compiler must have access to the source-code files gdt.f90 (containing the Global Data Module), typ.f90 (Datatypes Module), and utl.f90 (Utilities Module).

```
PROGRAM INPUT_BLOCK_EXAMPLE
USE DATATYPES; USE UTILITIES
INTEGER :: IERR, KSTATES = 0, MORE
TYPE (LLIST), POINTER :: BLOCKPTR, TAILPTR
CHARACTER(LEN=40), DIMENSION(0)  :: NOCOL
CHARACTER(LEN=12), DIMENSION(50) :: STNAME
INTEGER,           DIMENSION(50) :: SHYEAR, POP
CHARACTER(LEN=8),  DIMENSION(50) :: OCEAN, STCAP
LOGICAL, DIMENSION(50) :: ORIG13
50 FORMAT('State name  ',2X,'Capital ',2X,'Year',2X,'Population',3X, &
          'Ocean',4X,'Original?',/,'------------',2X,'--------',2X,  &
          '----',2X,'----------',2X,'--------',2X,'---------')
100 FORMAT(A,2X,A,2X,I4,2X,I9,3X,A,6X,L1)
OPEN(7,FILE='block.in',STATUS='OLD')
OPEN(8,FILE='block.out',STATUS='REPLACE')
OCEAN = 'no ocean'
CALL UTL_READBLOCK(0,'STATE_DATA',NOCOL,7,8,'STATENAME',   &
                    .TRUE.,BLOCKPTR,TAILPTR,KSTATES)
CALL UTL_FILTERLIST(BLOCKPTR,8,'STATENAME',KSTATES,IERR,STNAME,MORE)
CALL UTL_FILTERLIST(BLOCKPTR,8,'CAPITAL',KSTATES,IERR,STCAP,MORE)
CALL UTL_FILTERLIST(BLOCKPTR,8,'STATEHOODYEAR',KSTATES,IERR,SHYEAR,MORE)
CALL UTL_FILTERLIST(BLOCKPTR,8,'POPULATION',KSTATES,IERR,POP,MORE)
CALL UTL_FILTERLIST(BLOCKPTR,8,'OCEAN',KSTATES,IERR,OCEAN,MORE)
CALL UTL_FILTERLIST(BLOCKPTR,8,'ORIGINAL13',KSTATES,IERR,ORIG13,MORE)
WRITE(8,50)
DO I=1,KSTATES
  WRITE(8,100)STNAME(I),STCAP(I),SHYEAR(I),POP(I),OCEAN(I),ORIG13(I)
ENDDO
END PROGRAM INPUT_BLOCK_EXAMPLE
```

This program produces the following output to file block.out:

```
State name   Capital   Year  Population  Ocean     Original?
------------ --------  ----  ----------  --------  ---------
Colorado     Denver    1876    4301261   no ocean     F
New Jersey   Trenton   1787    8414350   Atlantic     T
```

Matrix Data Types

In some situations, such as defining a weight matrix for groups of observations within which errors are correlated, an array of matrices is needed. The dimensions of each matrix, or even the need to actually store data in the matrix, may vary from matrix to matrix. In these situations, efficient management of memory use can best be achieved by defining an array of a derived data type, where each element of the array is a structure containing a matrix that can be dynamically allocated to the required dimensions. The DMATRIX data type is designed for this purpose as follows:

```
TYPE :: DMATRIX
  CHARACTER(LEN=12)                          :: ARRAYNAME
  INTEGER                                    :: NR
  INTEGER                                    :: NC
  DOUBLE PRECISION, POINTER, DIMENSION(:,:)  :: DVAL
END TYPE DMATRIX
```

where components of the DMATRIX data type are as follows:
    ARRAYNAME is an optional name associated with the stored matrix.
    NR is the number of rows in the matrix.
    NC is the number of columns in the matrix.
    DVAL contains the matrix values in a DMATRIX structure.
One structure of the DMATRIX type can hold a 2-D DOUBLE PRECISION array dimensioned (NR, NC).

    The CDMATRIX data type is designed for compressed storage of matrix data and is defined as follows:

```
TYPE :: CDMATRIX
  CHARACTER(LEN=12)                      :: ARRAYNAME
  INTEGER                                :: IDIM
  INTEGER                                :: NNZ
  INTEGER                                :: NR
  INTEGER                                :: NC
  DOUBLE PRECISION, POINTER, DIMENSION(:) :: DVAL
  INTEGER,          POINTER, DIMENSION(:) :: IPOS
  INTEGER,          POINTER, DIMENSION(:) :: ICOL
END TYPE CDMATRIX
```

where the components of the CDMATRIX data type are as follows:
    ARRAYNAME is an optional name associated with the stored matrix.
    IDIM is the dimension of DVAL and IPOS.
    NNZ is the number of non-zero entries in the true matrix and in DVAL; it must not exceed IDIM.
    NR is the number of rows in the true matrix.
    NC is the number of columns in the true matrix, and the dimension of ICOL.
    DVAL contains non-zero elements of the true matrix, in column-major order.
    IPOS contains the position in the true matrix of the corresponding element in DVAL (assuming column-major order).
    ICOL contains the position in DVAL and IPOS of the first non-zero element in each column of the true matrix. For a column containing all zeros, the corresponding ICOL element equals 0.

    One CDMATRIX structure can hold a compressed, 2-D DOUBLE PRECISION array dimensioned (NR, NC).

    Compressed matrices are stored in column-major order. In column-major storage order, all entries of column 1 are numbered first, starting at row 1, followed by all entries of column 2, and so on. For example, for a sparse matrix of six rows and eight columns, with column-major storage order the columns in the matrix are numbered as follows:

1	7	13	19	25	31	37	43
2	8	14	20	26	32	38	44
3	9	15	21	27	33	39	45
4	10	16	22	28	34	40	46
5	11	17	23	29	35	41	47
6	12	18	24	30	36	42	48

31

With column-major ordering, the resulting values of DVAL and IPOS are consistent with the order used by Fortran. That is, if a CDMATRIX structure were used to hold a matrix populated in all elements with non-zero values, the DVAL array would be identical to a standard 2-D array of DOUBLE PRECISION type.

To illustrate, assume a sparse matrix of six rows and eight columns is to be stored in a CDMATRIX structure. Assume that the matrix is populated as follows:

11.	0.	0.	0.	0.	0.	0.	0.
21.	0.	0.	0.	0.	0.	0.	0.
0.	32.	33.	0.	0.	35.	0.	0.
0.	0.	0.	44.	45.	0.	0.	0.
0.	0.	53.	0.	0.	0.	0.	58.
0.	0.	0.	0.	0.	0.	67.	0.

This matrix contains 10 non-zero values; therefore, NNZ = 10. This matrix can be stored in a CDMATRIX structure in compressed form as:

```
IDIM = 10
NNZ = 10
NR = 6
NC = 8
DVAL = 11., 21., 32., 33., 53., 44., 45., 35., 67., 58.
IPOS = 1, 2, 9, 15, 17, 22, 28, 33, 42, 47
ICOL = 1, 3, 4, 6, 7, 8, 9, 10
```

To avoid problems when using the CDMATRIX data type, it is best to call subroutine TYP_NULL to nullify pointers in each CDMATRIX structure when it is declared.

Character-String Data Type

In some situations it is most efficient to have a CHARACTER array in which each element of the array contains an array that can be dynamically allocated to the required dimension. The CHAR_1D_30 data type supports this need, and is defined as follows:

```
TYPE :: CHAR_1D_30
  INTEGER                                :: NDIM
  CHARACTER(LEN=30), POINTER, DIMENSION(:) :: CVAL
END TYPE CHAR_1D_30
```

One structure of the CHAR_1D_30 type can hold a 1-D CHARACTER (LEN=30) array dimensioned (NDIM).

Chapter 4: Datatypes Module

Public Subprograms

The public subprograms of the Datatypes Module are listed in Table 4-1 and additional information is provided in Appendix B.

Table 4-1. Public subprograms of the Datatypes Module.

Subprogram name	Purpose
TYP_DEALLOC	Deallocate all arrays and nullify all pointers in a structure of a derived data type.
TYP_NULL	Nullify all pointers in a structure of a derived data type.

Chapter 5: Global Data Module
By Edward R. Banta and John E. Doherty

Data contained in the Global Data Module are considered global and are made available to any program unit by including the statement "USE GLOBAL_DATA". Data of the Global Data Module are documented in Appendix C. The source code for the Global Data Module is in a file named gdt.f90.

The data in the Global Data Module include the following.

1. The version number of the JUPITER API.

2. Extreme default values used to clearly show that variables have not been assigned values by the application.

3. The number of characters allowed for the names of observations or predictions.

4. The number of characters available for long strings, such as those containing pathnames or command lines.

5. A value used to format data-exchange files when there are many hundreds of defined parameters.

6. Strings of blanks and hyphens that can be used in creating output files,

7. Support for error messages.

8. A variable that can be used to control the detail printed to output files.

Chapter 6: Utilities Module
By Edward R. Banta, John E. Doherty, and Eileen P. Poeter

The Utilities Module contains subprograms that generally are useful to multiple modules or are expected to be useful in applications. Most subprograms in the Utilities Module are public. No public data are stored in the Utilities Module. The Utilities Module contains private subprograms and data that are not documented in this report. The source code for the Utilities Module is in a file named utl.f90. The Utilities Module uses the Datatypes and Global Data Modules, which need to be available when utl.f90 is compiled.

One responsibility of the Utilities Module is to open and close data-exchange files. The files and associated file-name extensions are described in Chapter 17.

Public Subprograms

The public subprograms of the Utilities Module are listed in Table 6-1; detailed descriptions of the subprograms are in Appendix D. The utility subprograms are categorized by primary purpose. The purposes are data-exchange file management, data input, data output, data manipulation, and error processing.

Table 6-1. Public subprograms of the Utilities Module.
[If the utility is primarily used by one aspect of the API, the relevant chapter or section is listed.]

Subprogram name	Purpose
Data-Exchange File Management (Chapter 17)	
UTL_DX_CLOSE	Close a data-exchange file.
UTL_DX_GETUNIT	Return unit number associated with a data-exchange file.
UTL_DX_OPEN	Open a data-exchange file and return the unit number for it.
Data Input	
UTL_DX_READ_MCMV	Read a "_mc" or "_mv" data-exchange file (Chapter 17).
UTL_DX_READ_WT	Read a "_wt" or "_wtpri" data-exchange file (Chapter 17).
UTL_GETARG	Return a command-line argument.
UTL_GETLINE	Return the next non-blank, non-comment line read from a file.
UTL_READ2D	Read a 2-D array from a file.
UTL_READBLOCK	Read an input block (Chapter 3) from a file.
UTL_READMATRIX	Read a standard or compressed matrix (Appendix B) from a file.

Subprogram name	Purpose
Data Output	
UTL_COLLBL	Write strings as labels for columns of a matrix.
UTL_COLNO	Write numbers as labels for columns of a matrix.
UTL_DX_WRITE_MCMV	Write a "_mc" or "_mv" data-exchange file (Chapter 17).
UTL_DX_WRITE_PA	Write a "_pa" data-exchange file (Chapter 17).
UTL_DX_WRITE_PCC	Write a "_pcc" data-exchange file (Chapter 17).
UTL_DX_WRITE_WT	Write a "_wt" or "_wtpri" data-exchange file (Chapter 17).
UTL_EIGEN	Calculate eigenvalues and eigenvectors of a variance-covariance matrix.
UTL_PARSQMATRIX	Write a square matrix labeled with parameter names.
UTL_TABLESECTIONS	Return number of table sections required to write a table.
UTL_WRITE2D	Write a 2-D array.
UTL_WRITEBLOCK	Write data read from an input block (Chapter 3).
UTL_WRITECDMATRIX	Write the contents of a CDMATRIX structure (Chapter 4).
UTL_WRITEMATRIX	Write a matrix with column and row labels.
UTL_WRITE_MESSAGE	Write the contents of AMESSAGE of the Global Data Module (Chapter 5).
UTL_WRTSIG	Write a number to a character value with maximum precision.
Data Manipulation – Statistical	
UTL_CALCWT	Calculate weight and variance for a value using a statistic of a specified type.
UTL_CHISQ	Determine Chi-squared statistic for N degrees of freedom for a two-sided significance level of 5 percent.
UTL_FSTT	Determine the value of the F statistic for a significance level of 5 percent.
UTL_STUD_T	Determine the value of the Student's t statistic for a two-sided significance level of 5 percent.

Subprogram name	Purpose
Data Manipulation – Character and string management	
UTL_ADDQUOTE	Enclose a string in double quotes if it includes one or more embedded spaces.
UTL_ARR2STRING	Convert an array of characters to a character string.
UTL_CASE	Copy a character string to a new variable in all uppercase or all lowercase.
UTL_CASETRANS	Convert a character string to all uppercase or all lowercase.
UTL_CHAR2NUM	Convert a character string to a number.
UTL_CHECK_NAMES	Check an array of names for conformance with the JUPITER naming convention (Chapter 2).
UTL_COMPRESSLINE	Remove excess blank characters from a character string.
UTL_COUNTSUBS	Return the number of occurrences of a substring in a string.
UTL_GETTOKEN	Return a string, starting at a specified position in another string.
UTL_GETWORD	Find the start and end of a space-delimited string, starting at a specified position in another string.
UTL_GETWORDLEFT	Return a space-delimited string, starting at a specified position in another string.
UTL_NUM2CHAR	Return the character equivalent of a number.
UTL_REMCHAR	Replace each occurrence in a string of a specified substring with blanks.
UTL_RWORD	Extract a string from another string, and optionally convert the word to a number.
UTL_SAMENAME	Make a case-insensitive comparison of two strings.
UTL_SHELLSORT	Sort an array of character strings in ascending order.
UTL_STRING2ARR	Store a character string in an array of characters.
UTL_TABREP	Replace tab characters in a string with blanks.
UTL_WHICH1	Find a specified string in an array of strings.

Chapter 6: Utilities Module

Subprogram name	Purpose
Data Manipulation – Matrices (Chapter 4, section "Matrix Data Types")	
UTL_ARR2CDMATRIX	Convert a 2-D array to a structure of derived type CDMATRIX .
UTL_COMBINESQMATRIX	Combine two square matrices corner-to-corner to create a third—all of type CDMATRIX.
UTL_CONSTRUCTCDMATRIX	Allocate and initialize a CDMATRIX structure to hold a specified number of nonzero values.
UTL_CONSTRUCTDMATRIX	Allocate and initialize a DMATRIX structure.
UTL_DIAGONAL	Extract diagonal elements from a square CDMATRIX structure.
UTL_GETCOL	Populate a 1-D array with data from one column of a compressed matrix.
UTL_GETICOL	Return column index for a specified element in a compressed-matrix data structure.
UTL_GETIROW	Return row index for a specified element in a compressed-matrix data structure.
UTL_GETPOS	Return position in a compressed-matrix data structure for a specified row and column.
UTL_GETROW	Populate a 1-D array with data from one row of a compressed matrix.
UTL_GETVAL	Return a specified element in a compressed-matrix data structure.
UTL_MATMUL	Return the result of multiplication of two matrices.
UTL_MATMULVEC	Return the result of multiplication of a matrix times a vector.
UTL_PREPINVERSE	Prepare a CDMATRIX structure suitable for containing the inverse of the input CDMATRIX structure.
UTL_SVD	Calculate the inverse and square root of the inverse of a square matrix stored in a CDMATRIX structure.
UTL_VEC2CDMATRIX	Convert a 1-D array to a CDMATRIX structure with the diagonal elements of the matrix populated by values from the array.
Data Manipulation – Input blocks (Chapter 4, section "Input-Block Data Types")	
UTL_APPENDLIST	Append one input-block data structure to another.
UTL_FILTER	Extract a scalar value from an input-block data structure.
UTL_FILTERLIST	Populate an array from an input-block data structure.
UTL_GROUPLIST	Insert data defined for groups in one input-block data structure into a second input-block data structure to serve as default value(s), where group names match.
UTL_MERGELIST	Merge two input-block data structures by matching key items.

Subprogram name	Purpose
Data Manipulation – Other	
ASSIGNMENT(=)	Overload the "=" operator, to assign a value to a structure of a derived type.
UTL_COVMAT	Build a combined variance-covariance matrix from a set of group variance-covariance matrices and an array of variances.
UTL_ELAPSED_TIME	Calculate elapsed time from an earlier time to current time.
UTL_ENDTIME	Write current time and elapsed time, relative to an earlier time.
UTL_GETUNIT	Return an unused unit number between two specified integers.
UTL_NEXTUNIT	Return the first unused unit number between 10 and 1000.
UTL_SUBSTITUTE	Populate an array with parameter values, using arrays containing values of adjustable and non-adjustable parameters.
UTL_SYSTEM	Submit a command to the operating system.
Error Processing	
UTL_STOP	Stop program execution; optionally write a message.
UTL_SUBERROR	Write a "programming error" message and stop program execution.
UTL_VERSION _CHECK	Check that JUPITER API version meets a specified version requirement.

Chapter 7: Basic Module
By Edward R. Banta and John E. Doherty

The Basic Module contains data and subprograms related to tasks that are commonly needed in any model-analysis application. Data and subprograms of the Basic Module are documented in Appendix E. The source code for the Basic Module is in the bas.f90 file. The Basic Module uses the Datatypes, Global Data, and Utilities Modules, which need to be available when bas.f90 is compiled.

The Basic Module includes subroutines for setting application options, defining a set of system commands to be used to invoke model runs, assigning the system command to be used when a forward model run is required, and invoking model runs. The module also can read matrix data to define a set of variance-covariance matrices that may be required by a JUPITER application.

Public Subprograms

The public subprograms of the Datatypes Module are listed in Table 7-1; detailed descriptions of the subprograms are provided in Appendix E.

Table 7-1. Public subprograms of the Basic Module.
[Subprograms are listed in the order in which they typically would be called. Subprogram naming conventions are described in Table 2-3]

Subprogram name	Purpose
BAS_INI_GETOPTIONS	Read an input block (Chapter 3) labeled "OPTIONS" and define selected variables.
BAS_INI_MODELEXEC	Read an input block labeled "MODEL_COMMAND_LINES", allocate arrays related to commands to be used to initiate model runs, and store the commands and related data.
BAS_INI_COVMAT	Read and store all required variance-covariance matrices.
BAS_GEN	Populate an array with parameter values, ensuring that they equal the values that will be written to model-input files.
BAS_EXE_SELECT	Assign integer identifying which command is to be executed to make a forward model run; requires previous call to BAS_INI_MODELEXEC.
BAS_EXE	Initiate a model run by submitting a specified command to the operating system; requires previous call to BAS_INI_MODELEXEC.
BAS_CLN	Deallocate all arrays and pointers of the module.

Input Blocks Read by the Basic Module

Three input blocks may be read by the Basic Module (Table 7-2). The OPTIONS block contains settings for variables used by any module, including modules that do not read input blocks of their own. The MODEL_COMMAND_LINES input block contains information related to commands to be submitted to the operating system to initiate a run of a model that is being analyzed. The MATRIX_FILES input block contains information for files that contain variance-covariance matrices for groups of observations or for groups of prior-information equations.

Table 7-2. Input blocks read by the Basic Module.

Input block	LLIST pointer	Read by subroutine	Required?[1]
OPTIONS	LLPTROPT	BAS_INI_GETOPTIONS	No
MODEL_COMMAND_LINES	LLPTRMODCOM	BAS_INI_MODELEXEC	Yes
MATRIX_FILES	LLPTRMATFIL	BAS_INI_COVMAT	Yes, but see footnote[2]

[1] No, the REQUIRED argument in the call to UTL_READBLOCK is false so that the input block is optional. Yes, the REQUIRED argument is true and the input block needs to appear in the file open on the unit identified by the INUNIT argument in the call to UTL_READBLOCK.

[2] The REQUIRED argument in the call to UTL_READBLOCK is true, but the call to UTL_READBLOCK is made only if the NCOVMAT argument of BAS_INI_COVMAT (Appendix E) is greater than zero.

The OPTIONS input block (Table 7-3) provides a mechanism for setting options that affect multiple modules of the JUPITER API or the application itself. The verbosity of messages and warnings that may be generated by some program units of the API is controlled by the IVERB variable of the Global Data Module (Chapter 5; also see Appendix C for the meanings of specific values of IVERB). In the OPTIONS input block, the keyword "VERBOSE" is used to assign a value to IVERB. The name of a derivatives interface file (Appendix A) is defined using the keyword "DERIVATIVES_INTERFACE".

Table 7-3. Data of the OPTIONS input block.
[No default column order is defined for the OPTIONS input block. Table entries include all data recognized by the Basic Module. "len" is the maximum number of characters in the string.]

Keyword or column name	Data type	Default	Description
VERBOSE	Integer	3	Controls verbosity of output. Value is assigned to module variable IVERB.
DERIVATIVES_INTERFACE	Character (len=MAX_STRING_LEN[1])	No default	Name of derivatives interface file (Appendix A). Value is assigned to module variable DERIV_INTERFACE.

[1] See Chapter 5 and Appendix C.

The MODEL_COMMAND_LINES input block (Table 7-4) is used to define a set of model commands and, for each, a purpose and a command identifier. Each system command for making a model run (or "model command") has a purpose as defined either by default or by application input. The purposes that can be assigned to model commands include "FORWARD", "FORWARD&DER", and "DERIVATIVES". "FORWARD" indicates the model command runs the model to generate model-calculated dependent values. "FORWARD&DER" indicates the model command runs the model to generate model-calculated dependent values and model-calculated derivatives of dependent values with respect to one or more parameters, and "DERIVATIVES" indicates the model command runs the model to generate model-calculated derivatives of dependent values with respect to one or more parameters. Note that the "FORWARD&DER" and "DERIVATIVES" purposes are only applicable when the model (for example, MODFLOW-2000, Hill and others, 2000) is capable of calculating sensitivities of dependents (observations and [or] predictions) with respect to parameters (see "Derivatives Interface File" section of Appendix A). Each model command also is assigned a command identifier, which is used in programming to distinguish among model commands when multiple model commands are defined.

Table 7-4. Data of the MODEL_COMMAND_LINES input block.
[Column names are in default column order (Chapter 3) and include all data recognized by the Basic Module. "len" is the maximum number of characters in the string.]

Keyword or column name	Data type	Default	Description
COMMAND[1]	Character (len=MAX_STRING_LEN[2])	No default	Command to be submitted to operating system to initiate a model run.
PURPOSE	Character (len=12)	"FORWARD"	Purpose of COMMAND. Supported options are "FORWARD", "FORWARD&DER", "DERIVATIVES"[3]
COMMANDID	CHARACTER (len=12)	No default	Identifier for COMMAND.

[1] COMMAND is the keyitem (Chapter 3) for the MODEL_COMMAND_LINES input block.
[2] See Chapter 5 and Appendix C.
[3] "FORWARD", run process model to generate model-calculated dependent values.
"FORWARD&DER", run process model to generate model-calculated dependent values and model-calculated derivatives of dependent values with respect to one or more parameters.
"DERIVATIVES", run process model to generate model-calculated derivatives of dependent values with respect to one or more parameters.

The MATRIX_FILES input block (Table 7-5) is intended to define names of files from which variance-covariance matrices are to be read. Each file name is recognized by the keyword "MATRIXFILE", and the number of matrices to be read from each file is defined by the value associated with the keyword "NMATRICES". When subroutine BAS_INI_COVMAT is called, it expects to read an input block labeled MATRIX_FILES. BAS_INI_COVMAT reads the input block, it opens the files named in the input block, and it reads the contents of those files to populate the COVMATARR array (Appendix E) of type CDMATRIX (Chapter 4).

Table 7-5. Data of the MATRIX_FILES input block.
[Table entries include all data recognized by the Basic Module. "len" is the maximum number of characters in the string.]

Keyword or column name	Data type	Default	Description
MATRIXFILE[1]	Character (len=MAX_STRING_LEN[2])	No default	File name or pathname containing matrix data in Complete Matrix or Compressed Matrix format (Chapter 3 and Appendix A).
NMATRICES	Integer	1	Number of matrices to be read from file MATRIXFILE.

[1] MATRIXFILE is the keyitem (Chapter 3) for the MATRIX_FILES input block.
[2] See Chapter 5 and Appendix C.

Chapter 8: Model Input-Output Module
By John E. Doherty and Edward R. Banta

The Model Input-Output Module contains subprograms for implementing communication between a JUPITER application and one or more models. It supports dependents (Chapter 10). This module also includes a small amount of public data. The Model Input-Output Module is accessed by the statement "USE MODEL_IO". Data and subprograms of the Model Input-Output Module are documented in Appendix F. The source code for the Model Input-Output Module is in the mio.f90 file. The Model Input-Output Module uses the Datatypes, Global Data, and Utilities Modules (Table 2-4), which need to be available when mio.f90 is compiled.

The Model Input-Output Module can distinguish among multiple categories of model-calculated values to be extracted from model-output files. Only one category of data can be extracted from a model-input file each time it is opened, but one file can be opened multiple times. The programmer determines the number of categories to be supported by an application. The module supports up to nine categories, each identified by a string of up to six characters. The number of categories required by an application is rarely expected to exceed two. The end user defines a category for each model-output file listed in the Model_Output_Files block using keyword CATEGORY. The API documented in this work supports two categories: (1) simulated equivalents to observations (CATEGORY=OBS) and (2) predictions (CATEGORY=PRED). If an application supports only a single category, the CATEGORY keyword in the Model_Output_Files block is optional.

Extraction of model-calculated values is performed by the MIO_EXT subroutine. During any given invocation of MIO_EXT, any or all of the categories of model-calculated values can be extracted from model-output files. The list of strings identifying all application-supported categories is provided to MIO_EXT in the MCVCAT argument. The categories for which values are to be extracted are provided to MIO_EXT in the MCVUSE argument. If an element of MCVUSE is true, the corresponding category in MCVCAT is used. When MIO_EXT is called, it extracts values from each model-output file with a category for which MCVUSE is true.

Public Subprograms

The public subprograms of the Model Input-Output Module are listed in Table 8-1; detailed descriptions of the subprograms are in Appendix F.

Table 8-1. Public subprograms of the Model Input-Output Module.
[Subprograms are listed in the order in which they typically would be called. Subprogram naming conventions are described in Table 2-3]

Subprogram name	Purpose
Subprograms that serve the initialize (INI) task	
MIO_INI_ALLOC	Initialize the module and allocate an array.
MIO_INI_INPUTFILES	Read an input block labeled "MODEL_INPUT_FILES", allocate arrays, and populate two of the arrays with data from the input block.
MIO_INI_OUTPUTFILES	Read an input block labeled "MODEL_OUTPUT_FILES", allocate arrays, and populate three of the arrays with data from the input block.
MIO_INI_TEMPLATE	Check template files and, for each parameter, find the smallest substitution width. The width is used to ensure consistent number of significant digits is used to write the value of each parameter. Requires previous call to MIO_INI_INPUTFILES or MIO_INI_INPUTFILES_RUNNER.
MIO_INI_INSTRUCT1	Read extraction-instruction files to determine dimensioning variables; requires previous call to MIO_INI_OUTPUTFILES or MIO_INI_OUTPUTFILES_RUNNER.
MIO_INI_INSTRUCTALLOC	Allocate arrays for storing extraction instructions; requires previous call to MIO_INI_INSTRUCT1 or MIO_INI_INSTRUCT_RUNNER.
MIO_INI_INSTRUCT2	Read extraction-instruction files and store instructions; requires previous call to MIO_INI_OUTPUTFILES or MIO_INI_OUTPUTFILES_RUNNER.
MIO_INI_DIMENSION	Assign dimensioning integers used in the module; requires previous call to MIO_INI_INSTRUCT1.
MIO_INI_ARRAYS	Populate argument-list arrays with data stored in arrays of the module; requires previous call to MIO_INI_INPUTFILES or MIO_INI_INPUTFILES_RUNNER.
Subprograms that serve the initialize (INI) task as needed for Parallel Processing	
MIO_INI_INPUTFILES_RUNNER	Allocate arrays and populate two of the arrays with data in the argument list—to be called by runner program.
MIO_INI_OUTPUTFILES_RUNNER	Assign module dimensioning variables, dimension arrays, and populate module arrays with data in the argument list related to model-output files—to be called by runner program.
MIO_INI_INSTRUCT_RUNNER	Assign module dimensioning variables, dimension arrays, and populate module arrays with data in the argument list related to extraction-instruction files—to be called by runner program.

Chapter 8: Model Input-Output Module

Subprogram name	Purpose
Subprograms that serve the Extract-model-calculated-values (EXT) task	
MIO_EXT_MODOUTFILE_CHANGE	Store a model_output file name; requires previous call to MIO_INI_OUTPUTFILES or MIO_INI_OUTPUTFILES_RUNNER.
MIO_EXT	Read model-output files and extract and store model-calculated values; requires previous calls to MIO_INI_OUTPUTFILES or MIO_INI_OUTPUTFILES_RUNNER, and MIO_INI_INSTRUCT1.
Subprograms that serve the Cleanup (CLN) task	
MIO_CLN_OUTFILES	Delete all model-output files.
MIO_CLN_DEALLOC	Deallocates all arrays of the module.

Input Blocks Read by the Model Input-Output Module

Two input blocks can be read by the Model-Input-Output Module (Table 8-2). The MODEL_INPUT_FILES input block (Table 8-3) contains information related to the files that are to be prepared by the Model Input-Output Module and used as input by a model being analyzed. The MODEL_OUTPUT_FILES input block (Table 8-4) contains information related to the files produced by a model being analyzed and to the reading of model-calculated values from those files.

Table 8-2. Input blocks read by the Model Input-Output Module.

Input block	LLIST pointer	Read by subroutine	Required?[1]
MODEL_INPUT_FILES	LLPTRMODIN	MIO_INI_INPUTFILES	Yes
MODEL_OUTPUT_FILES	LLPTRMODOUT	MIO_INI_OUTPUTFILES	Yes

[1] Yes, the REQUIRED argument in the call to UTL_READBLOCK is true and the input block needs to appear in the file that is open on the unit identified by the INUNIT argument in the call to UTL_READBLOCK.

Table 8-3. Data of the MODEL_INPUT_FILES input block.
[Column names are in default column order (Chapter 3) and include all data recognized by the MODEL Input-Output Module. "len" is the maximum number of characters in the string.]

Keyword or column name	Data type	Default	Description
MODINFILE[1]	Character (len=MAX_STRING_LEN[2])	No default	Name or pathname of a model-input file.
TEMPLATEFILE	Character (len=MAX_STRING_LEN[2])	No default	Name or pathname of a template file for preparing file MODINFILE.

[1] MODINFILE is the keyitem (Chapter 3) for the MODEL_INPUT_FILES input block.
[2] See Chapter 5 and Appendix C.

Table 8-4. Data of the MODEL_OUTPUT_FILES input block.
[Column names are in default column order (Chapter 3) and include all data recognized by the MODEL Input-Output Module. "len" is the maximum number of characters in the string.]

Keyword or column name	Data type	Default	Description
MODOUTFILE[1]	Character (len=MAX_STRING_LEN[2])	No default	Name or pathname of a model-output file.
INSTRUCTIONFILE	Character (len=MAX_STRING_LEN[2])	No default	Name or pathname of a file containing extraction instructions for file MODOUTFILE.
CATEGORY	Character (len=6)	"0"	Category of dependents to be extracted from file MODOUTFILE. Supported options are defined by argument MCVCAT of subroutine MIO_INI_INSTRUCT2.

[1] MODOUTFILE is the keyitem (Chapter 3) for the MODEL_OUTPUT_FILES input block.
[2] See Chapter 5 and Appendix C.

Chapter 9: Equation Module
By John E. Doherty

The Equation Module provides the ability to parse and evaluate mathematical expressions supplied as lines of text. These expressions can include named variables, the values of which can change with time. They can have logical or numerical outcomes. Thus they can be used as the basis for decision making, for the provision of linear and nonlinear prior information, for the calculation of "derived parameters" from "primary parameters," and for the calculation of "derived dependents" from "primary dependents" read from model-output files.

A mathematical expression (referred to as an "equation" herein) is supplied as a line of text, the syntax of which needs to follow Fortran coding conventions, except that variable names in equations must not contain uppercase letters. The equation text cannot include an "=" sign, because the assignment of calculated values to variables takes place outside the Equation Module. Each equation needs to be supplied with a name. Errors in equations are written to the AMESSAGE character string made public by the Global Data Module. An error condition is identified by a non-zero value for the IFAIL variable returned by most Equation Module subroutines.

The Equation Module distinguishes between two types of equations: standard equations and linear prior-information equations.

Standard equations are parsed and stored; evaluation takes place when needed.

Linear prior-information equations are not stored literally by the Equation Module. Instead, the Equation Module calculates and stores the "constant term" (the value obtained when all variables are set to zero) and the linear coefficients of all variables used by the equation. If parameters are log-transformed, the log of the parameter needs to appear in the equation. If the Equation Module detects that an equation is not linear, it returns an error flag and error message.

Data and public subprograms of the Equation Module are documented in Appendix G. Source code for the Equation Module resides in a file named eqn.f90. The Equation Module uses the Global Data and Utilities Modules, which need to be available when eqn.f90 is compiled.

Public Subprograms

The public subprograms of the Datatypes Module are listed in Table 9-1; detailed descriptions of the subprograms are in Appendix G.

Table 9-1. Public subprograms of the Equation Module.
[Subprograms are listed in order in which they typically would be called. Subprogram naming conventions are described in Table 2-3]

Subprogram name	Purpose
EQN_INI	Allocate memory and initialize module variables and arrays.
EQN_INI_INSTALL	Store an equation; requires previous call to EQN_INI.
EQN_LINEAR_COEFFS	Parse a linear equation and return coefficients and constant term.
EQN_GETNAME	Return the name of an equation corresponding to a specified equation number; requires previous calls to EQN_INI and EQN_INI_INSTALL.
EQN_EVALUATE	Evaluate an equation, using variable values substituted for variable names.
EQN_CLN	Deallocate memory allocated by the module.

Functionality

The functionality of the Equation Module includes arithmetic and logical operators and functions.

Arithmetic Operators

Arithmetic operators implemented in the Equation Module are listed in Table 9-2. The order in which mathematical operations are carried out in evaluating a mathematical expression is the same as that used in normal mathematical operations. That is, raising a value to a power is followed by multiplication and division, followed by unary addition and subtraction, followed by binary addition and subtraction. Parentheses can be used to override this order.

Logical Operators

Logical operators are used in equations that have the result *.TRUE.* or *.FALSE.*. They are used in determining whether certain conditions are met. Logical operators available through the Equation Module are listed in Table 9-3. Use parentheses (Table 9-2) as needed to avoid ambiguity where order of evaluation of logical operators and arithmetic operators could be interpreted in more than one way.

Functions

A number of functions are implemented in the Equation Module. These functions are listed in Table 9-4.

Table 9-2. Arithmetic operators supported by the Equation Module.

Operator symbol	Operation
( )	<u>Brackets</u>. Terms within brackets are evaluated first. For example:- $5 + 4 * 3$ is evaluated as 17. However $(5 + 4) * 3$ is evaluated as 27.
** or ^	<u>Power</u>. $a**b$ or a^b is interpreted as "a raised to the power b".
/	<u>Division</u>. a/b is interpreted as "a divided by b".
*	<u>Multiplication</u>. $a*b$ is interpreted as "a multiplied by b".
-	<u>Subtraction</u>. This can be a unary or binary operator. $a-b$ is interpreted as "a minus b"; $-a$ is interpreted as "negative a".
+	<u>Addition</u>. This can be a unary or binary operator. $a+b$ is interpreted as "a plus b"; $+a$ is interpreted as "positive a".

Table 9-3. Logical operators supported by the Equation Module.

Operator symbol	Operation
.lt.	<u>Less than</u>. *a*.lt.*b* is *.TRUE.* if *a* is less than *b*.
.le.	<u>Less than or equal to</u>. *a*.le.*b* is *.TRUE.* if *a* is less than or equal to *b*.
.eq.	<u>Equal to</u>. *a*.eq.*b* is *.TRUE.* if *a* equals *b*.
.gt.	<u>Greater than</u>. *a*.gt.*b* is *.TRUE.* if *a* is greater than *b*.
.ge.	<u>Greater than or equal to</u>. *a*.ge.*b* is *.TRUE.* if *a* is greater than or equal to *b*.
.ne.	<u>Not equal to</u>. *a*.ne.*b* is *.TRUE.* if *a* does not equal *b*.
.and.	<u>And</u>. *a*.and.*b* is *.TRUE.* if both *a* and *b* are true; for example (1.lt.10).and.(6.lt.7) is *.TRUE.*
.or.	<u>Or</u>. *a*.or.*b* is *.TRUE.* if *a* is *.TRUE.* or *b* is *.TRUE.* or both are *.TRUE.*; for example, (1.lt.10).or.(1.lt.0) is *.TRUE.*

Table 9-4. Functions supported by the Equation Module.

Function	Definition
abs( )	<u>Absolute value</u>. Argument can be any floating-point number.
cos( )	<u>Cosine</u>. Argument can be any floating-point number supplied in radians.
acos( )	<u>Inverse cosine</u>. Absolute value of argument needs to be less than or equal to one. Value is returned in radians and lies in the range 0 to π.
sin( )	<u>Sine</u>. Argument can be any floating-point number supplied in radians.
asin( )	<u>Inverse sine</u>. Absolute value of argument needs to be less than or equal to one. Value is returned in radians and lies in the range $-\pi/2$ to $\pi/2$.
tan( )	<u>Tan</u>. Argument can be any floating-point number supplied in radians.
atan( )	<u>Inverse tan</u>. Argument can be any floating-point number. Value is returned in radians and lies in the range $-\pi/2$ to $\pi/2$.
cosh( )	<u>Hyperbolic cosine</u>. Argument can be any floating-point number.
sinh( )	<u>Hyperbolic sine</u>. Argument can be any floating-point number.
tanh( )	<u>Hyperbolic tan</u>. Argument can be any floating-point number.
exp( )	<u>Exponential</u>. Argument can be any floating-point number.
log( )	<u>Log to base *e*</u>. Argument needs to be a positive floating-point number.
Log10( )	<u>Log to base 10</u>. Argument needs to be a positive floating-point number.
sqrt( )	<u>Square root</u>. Argument needs to be non-negative.
min(, ,)	<u>Minimum of a series of numbers</u>. Arguments can be any floating-point numbers.
max(, ,)	<u>Maximum of a series of numbers</u>. Arguments can be any floating-point numbers.
mod(,)	<u>Remainder</u>. mod(*a*,*b*) is the remainder after *a* is divided by *b*.

Examples

The following are some examples of equations with outcomes that are numbers. It is assumed that values are available for variables *par1*, *par2*, *par3* and *par4*.

```
par1 + sqrt(par3*par1)
sqrt(abs(sin(par3/57.29)))
exp(3.0 * sqrt(par2/par4))
par1
1.0
```

Some examples of equations that have a logical outcome (that is, for which the outcome is either *.TRUE.* of *.FALSE.*) are as follows:

```
mod(itn,200).eq.0                 (.TRUE. if itn is a multiple of 200)
par5.lt.0.23                      (.TRUE. if par5 is less than 0.23)
(mod(itn,200).eq.0).and.( par5.lt.0.23)   (.TRUE. if the two expressions above
are .TRUE.)
```

Notes:
1. Spaces can be inserted between operators, variable names, parentheses, and so on. However, a variable name cannot include a space.
2. An equation entity that is not an operator or a function is first tested as a number. If it cannot be read as a number, it is assumed to be a variable. To avoid confusion, variable names cannot begin with a digit.
3. Linear prior-information equations require stricter formatting than indicated above; see the description of subroutine EQN_LINEAR_COEFFS (Appendix G).
4. If an illegal argument is supplied to any function (for example, if a negative number is provided as the argument to a log function), an error condition arises and the error is reported through the AMESSAGE string. Thus, there should be no possibility of encountering a compiler-generated run-time error that would result in unexpected cessation of execution.

Chapter 10: Dependents Module for Observations and Predictions
By Edward R. Banta and Eileen P. Poeter

The term "Dependents" comes from mathematics, where a dependent variable is the quantity calculated using an algebraic equation or as the solution of a differential equation. As defined, dependents encompass both observations and predictions.

The Dependents Module stores data and contains subprograms for handling dependents. The Dependents Module is accessed by the statement "USE DEPENDENTS". The source code for the Dependents Module is in the dep.f90 file. The Dependents Module uses the Datatypes, Global Data, Utilities, and Equation Modules, which need to be available when dep.f90 is compiled.

Defining Statistics Related to Dependents

Calculations that use observations and model-calculated equivalents to observations commonly require that weighting be defined for each observation. Weights often are determined based on an analysis of observation error (Hill and Tiedeman, in press). When the errors are assumed to be independent, weight commonly is defined as the inverse of variance of the observation error. To allow the use of statistics that make the most sense in a given situation, the Dependents Module can calculate a weight from a variance, standard deviation, or coefficient of variation. The keywords "STATISTIC" and "STATFLAG" are used in this and later chapters for variables that control how weights are calculated. STATISTIC is a floating-point number used to calculate the weight, and STATFLAG is a text string that indicates what STATISTIC is and how to calculate the weight. Table 10-1 lists valid values of STATFLAG, corresponding interpretations of STATISTIC, and formulas by which weight can be calculated.

Table 10-1. Calculation of weight, using STATISTIC and STATFLAG.
[VAL is a value used in addition to STATISTIC for calculating weight. In the context of observations, VAL is the observed value.]

STATFLAG	Interpretation of STATISTIC	Formula used to calculate weight
CV	Coefficient of variation	Weight = $1.0/(STATISTIC*VAL)^2$
SD	Standard deviation	Weight = $1.0/(STATISTIC)^2$
VAR	Variance	Weight = $1.0/STATISTIC$
WT	Weight	Weight = STATISTIC
SQRWT	Square root of the weight	Weight = $(STATISTIC)^2$

For predictions, similar information is needed to define prediction intervals (Hill and Tiedeman, in press), and the Dependents Module uses analogous formulas for that purpose. In the context of predictions, a variance is used to quantify the error that can be expected to be associated with a field measurement corresponding to the prediction. The Dependents Module recognizes the keywords MEASSTATFLAG and MEASSTATISTIC to identify input used to assign variance values. MEASSTATISTIC is a floating-point number used to calculate the variance, and MEASSTATFLAG is a text string that indicates what MEASSTATISTIC is and how to calculate the variance, as described in Table 10-2.

55

Table 10-2. Calculation of variance, using MEASSTATISTIC and MEASSTATFLAG. [VAL stands for a value used in addition to MEASSTATISTIC for calculating variance. In the context of predictions, VAL is a reference value.]

MEASSTATFLAG	Interpretation of MEASSTATISTIC	Formula used to calculate variance
CV	Coefficient of variation	$(MEASSTATISTIC*VAL)^2$
SD	Standard deviation	$(MEASSTATISTIC)^2$
VAR	Variance	MEASSTATISTIC

Derived Dependents

In some cases values produced by the process model may need to be combined or otherwise mathematically manipulated to generate the desired observations or predictions. The Dependents Module supports definition of derived dependents (derived observations and derived predictions) to handle these cases. The required expressions are stored, and mathematical manipulations are carried out, by the Equation Module (Chapter 9). Each derived dependent is defined by the presence in the appropriate input block of the keyword or column name EQUATION along with an expression that defines how the derived dependent is to be calculated. To implement and use derived dependents requires calling the appropriate subroutines of the Equation Module to initialize the module and store the equations, and then to evaluate the equations. Use of derived dependents in an application is demonstrated in UCODE_2005 (Poeter and others, 2005).

Public Subprograms

The public subprograms of the Dependents Module are listed in Table 10-3; detailed descriptions of the subprograms are in Appendix H.

Chapter 10: Dependents Module for Observations and Predictions

Table 10-3. Public subprograms of the Dependents Module.
[Within categories for basic function, subprograms are listed in the order in which they typically would be called or in alphabetical order, if a calling order is not obvious. Subprogram naming conventions are described in Table 2-3. Here, many subprogram names include UEV because they are generally used in the "Use extracted values task.]

Subprogram name	Purpose
Subprograms designed to serve the initialize (INI) and clean (CLN) tasks	
DEP_INI_READ	Read input blocks labeled "OBSERVATION_GROUPS" and "OBSERVATION_DATA". Allocate and populate arrays with data from these input blocks. Optionally, write the data.
DEP_INI_ALLOC	Allocate and initialize some arrays of the module. Requires previous call to DEP_INI_READ.
DEP_INI_STORE	Allocate and populate arrays for dependents. Define extracted and derived dependents. Create weight matrix for observations. Optionally, write the data. Requires previous call to DEP_INI_ALLOC. Previous call to EQN_INI is needed if there are derived dependents.
DEP_INI_NAMCHK	Allocate and initialize arrays related to dependents. Determine the numbers of used, derived, and extracted dependents. Requires previous call to DEP_INI_STORE.
DEP_CLN	Deallocate all arrays and pointers of the module.
Subprograms designed to perform calculations for the extract-model-calculated-values (EXT) and use-extracted-values (UEV) tasks	
DEP_EXT_DER	Calculate simulated equivalents to derived dependents. Requires previous calls to EQN_INI and DEP_INI_STORE.
DEP_UEV_RESIDUALS	Calculate residuals and weighted residuals, using a full weight matrix; requires previous call to DEP_INI_STORE.
Subprograms designed to write to data-exchange files (Chapter 17) for the use-extracted-values (UEV) task and for all tasks	
DEP_UEV_DX_WRITE_OS	Write "_os"
DEP_UEV_DX_WRITE_P	Write "_p"; requires previous call to DEP_INI_READ.
DEP_DX_WRITE_GM	Write "_gm"
DEP_UEV_DX_WRITE_R	Write "_r"
DEP_UEV_DX_WRITE_SS	Write "_ss"
DEP_UEV_DX_WRITE_W	Write "_w"
DEP_UEV_DX_WRITE_WS	Write "_ws"
DEP_UEV_DX_WRITE_WW	Write "_ww"
Subprograms designed to read from data-exchange files (Chapter 17) for the use-extracted-values (UEV) task	
DEP_UEV_DX_READ_OS	Read "_os"
DEP_UEV_DX_READ_WS	Read "_ws"
DEP_UEV_DX_READ_PNUM	Read number of predictions from "_p" data-exchange file.
DEP_UEV_DX_READ_P	Read "_p" data-exchange file

Subprogram name	Purpose
Other	
DEP_UEV_WRITEOBSTABLE	Write a table of observed and model-calculated values. Show residuals and weighted residuals. Requires previous call to DEP_INI_STORE.
DEP_GET_GROUP	Return group name and plot symbol value for specified dependent number. Requires previous call to DEP_INI_READ.
DEP_GET_PLOTSYMBOL	Return plot symbol value for specified group. Requires previous call to DEP_INI_READ.

Input Blocks Read by the Dependents Module

Input blocks that can be read by the Dependents Module are listed in Table 10-4.

Table 10-4. Input blocks read by the Dependents Module.

Input block	LLIST pointer	Read by subroutine	Required?[1]
Observations			
OBSERVATION_GROUPS	LLPTRGPOBS	DEP_INI_READ, when argument IDEPTYPE equals 1 or 3	No
OBSERVATION_DATA	LLPTROBS	DEP_INI_READ, when argument IDEPTYPE equals 1 or 3	Yes
DERIVED_OBSERVATIONS	LLPTROBS	DEP_INI_READ, when argument IDEPTYPE equals 1 or 3	No
Predictions			
PREDICTION_GROUPS	LLPTRGPOBS	DEP_INI_READ, when argument IDEPTYPE equals 2 or 3	No
PREDICTION_DATA	LLPTROBS	DEP_INI_READ, when argument IDEPTYPE equals 2 or 3	Yes
DERIVED_PREDICTIONS	LLPTROBS	DEP_INI_READ, when argument IDEPTYPE equals 2 or 3	No

[1]"No" means the REQUIRED argument in the call to UTL_READBLOCK is false; therefore, the input block is optional. "Yes" means that the REQUIRED argument is true and that the input block needs to appear in the file that is open on the unit identified by the INUNIT argument in the call to UTL_READBLOCK.

The OBSERVATION_GROUPS input block (Table 10-5) contains information related to groups of observations. The OBSERVATION_DATA and DERIVED_OBSERVATIONS input blocks (Table 10-6) contains information related to individual observations.

The PREDICTION_GROUPS input block (Table 10-7) contains information related to groups of predictions. The PREDICTION_DATA and DERIVED_ PREDICTIONS input blocks (Table 10-8) contain information related to individual predictions.

Table 10-5. Data of the OBSERVATION_GROUPS input block.
[No default column order is defined for the OBSERVATION_GROUPS input block. Table entries include all data recognized by the Dependents Module. "len" is the maximum number of characters in the string.]

Keyword or column name	Data type	Default	Description
GROUPNAME[1]	Character (len=12)	"DefaultObs"	Group name; values are stored in GROUPNAM array.
USEFLAG	Logical	True	True means include include group in analysis.
PLOTSYMBOL	Integer	1	Integer listed with group and group members in some data-exchange files (Chapter 17).
WTMULTIPLIER	Double precision	1.0	Value to be used to multiply weight of each group member.
COVMATRIX	Character (len=12)	No default	Name of variance-covariance matrix for group.
Other			Any keyword recognized by an associated Observations_Data input block.

[1] GROUPNAME is the keyitem (Chapter 3) for the OBSERVATION_GROUPS input block.

Table 10-6. Data of the OBSERVATION_DATA and DERIVED_OBSERVATIONS input blocks.
[Column names are in default column order (Chapter 3) and include all data recognized by the Dependents Module. "len" is the maximum number of characters in the string.]

Keyword or column name	Data type	Default	Description
OBSNAME[1]	Character (len=LENDNAM[2])	No default	Observation name; values are stored in the OBSNAM array.
OBSVALUE	Double precision	No default	Observed value.
STATISTIC	Double precision	No default	Statistic from which a weight is to be calculated (Table 10-1).
STATFLAG		No default	Flag indicating what STATISTIC is (Table 10-1).
GROUPNAME	Character (len=12)	"DefaultObs"	Group name.
EQUATION	Character (len=MAX_STRING_LEN[2])	No default	Equation defining how the equivalent simulated value will be calculated from simulated equivalents of previously defined observations.

[1] OBSNAME is the keyitem (Chapter 3) for the OBSERVATION_DATA input block.
[2] See Chapter 5 and Appendix C.

Table 10-7. Data of the PREDICTION_GROUPS input block.
[No default column order is defined for the PREDICTION_GROUPS input block. Table entries include all data recognized by the Dependents Module. "len" is the maximum number of characters in the string.]

Keyword or column name	Data type	Default	Description
GROUPNAME[1]	Character (len=12)	DefaultPred	Group name; values are stored in GROUPNAM array.
USEFLAG	Logical	True	True means include group in analysis.
PLOTSYMBOL	Integer	1	Integer to be associated with group and group members in some data-exchange files (Chapter 17).
Other			Any keyword from an associated Prediction_Data input block.

[1] GROUPNAME is the keyitem (Chapter 3) for the PREDICTION_GROUPS input block.

Table 10-8. Data of the PREDICTION_DATA and DERIVED_PREDICTIONS input blocks.
[Column names are in default column order (Chapter 3) and include all data recognized by the Dependents Module. "len" is the maximum number of characters in the string.]

Keyword or column name	Data type	Default	Description
PREDNAME[1]	Character (len=LENDNAM[2])	no default	Prediction name; values are stored in the OBSNAM array.
REFVALUE	Double precision	no default	Reference value; values are stored in the OBSVAL array.
MEASSTATISTIC	Double precision	no default	Statistic used to calculate the variance of the measurement error anticipated if the predicted item is measured in the field.
MEASSTATFLAG	Double precision	no default	Flag indicating what MEASSTATISTIC is (Table 10-2).
GROUPNAME	Character (len=12)	DefaultPred	Group name.
EQUATION	Character (len=MAX_STRING_LEN[2])	no default	Equation defining how the equivalent simulated prediction will be calculated from simulated values of previously defined predictions.

[1] PREDNAME is the keyitem (Chapter 3) for the PREDICTION_DATA input block.
[2] See Chapter 5 and Appendix C.

Chapter 11: Prior-Information Module
By Edward R. Banta and Eileen P. Poeter

It can be very useful to include measurements of parameter values in many situation. For example, in ground-water problems, aquifer tests can be used to obtain values of hydraulic conductivity. It can be advantageous to include these measurements in many types of model analysis. In addition, it can be useful to impose smoothness constraints on spatially varying parameters. Both situations are addressed using prior information.

The Prior-Information Module supports definition of prior information that can be expressed as a linear function of the parameters, which is most common. Thus, for the Prior-Information Module documented here the prior information needs to be of the form:

$$P'_p(\underline{b}) = \sum_{j=1}^{NP} (a_{p,j} b_j) = a_{p,1} b_1 + a_{p,2} b_2 + \ldots + a_{p,NPR} b_{NP}$$

where p indicates the pth prior-information equation, $a_{p,j}$ are coefficients, and b_j is the jth parameter value. Hill and Tiedeman (in press, Section 3.1.3) provide a discussion of prior information.

Data and subprograms of the Prior-Information Module are documented in Appendix I. The Prior-Information Module is accessed by the statement "USE PRIOR_INFORMATION". The source code for the Prior-Information Module is in a file named pri.f90. The Prior-Information Module uses the Datatypes, Global Data, and Equation Modules, which need to be available when pri.f90 is compiled.

Accounting for Prior-Information Error

Procedures that use prior information commonly require that weighting be defined for each prior-information equation. Weighting often is determined based on an analysis of error (Hill and Tiedeman, in press). When the errors are assumed to be independent, the weighting commonly is defined as the inverse of the variance of the prior-information error. To allow use of statistics that make the most sense to the user in a given situation, the Prior-Information Module can calculate a weight from a variance, standard deviation, or coefficient of variation. The names "STATISTIC" and "STATFLAG" are used for variables that control how weights are calculated. STATISTIC is a floating-point number used to calculate the weight, and STATFLAG is a text string that indicates what STATISTIC is and how to calculate the weight. The same options are available as for dependents, so Table 10-1 lists valid values of STATFLAG, corresponding interpretations of STATISTIC, and methods by which weights can be calculated for prior information. In the case of prior information, VAL is the measured parameter value. As in the Dependents Module (Chapter 10), the Prior-Information Module supports use of a variance-covariance matrix to define correlated errors in prior information.

Public Subprograms

The public subprograms of the Prior-Information Module are listed in Table 11-1; detailed descriptions of the subprograms are in Appendix I.

Table 11-1. Public subprograms of the Prior-Information Module.
[Subprograms are listed in the order in which they typically would be called. Subprogram naming conventions are described in Table 2-3]

Subprogram name	Purpose
PRI_INI_READ	Read input blocks labeled "PRIOR_INFORMATION_GROUPS" and "LINEAR_PRIOR_INFORMATION", allocate module arrays, and populate selected arrays.
PRI_INI_ALLOC	Allocate and initialize module arrays.
PRI_INI_STORE	Populate arrays with data from input-block data structures containing prior information; requires previous call to PRI_INI_READ.
PRI_INI_INSTALL	Store prior-information equations; requires previous calls to PRI_INI_STORE and EQN_INI.
PRI_INI_PROCESS	Check and store data related to prior information, and build a weight matrix for all prior information; requires previous call to PRI_INI_STORE.
PRI_INI_POPX	Populate an array of prior-information sensitivity; requires previous call to PRI_INI_PROCESS.
PRI_DEF	Define required number of model runs as 0.
PRI_UEV_RESIDUALS	Calculate residuals and weighted residuals for prior-information equations, using a full weight matrix; requires previous calls to PRI_INI_ALLOC, PRI_INI_INSTALL, and PRI_INI_PROCESS.
PRI_UEV_DX_WRITE_PR	Write "_pr" data-exchange file (Chapter 17); requires previous call to PRI_INI_STORE.
PRI_UEV_DX_READ_PR	Read "_pr" data-exchange file (Chapter 17).
PRI_CLN	Deallocate all arrays, pointers, and derived-type data structures of the module.

Input Blocks Read by the Prior-Information Module

Two input blocks can be read by the Prior-Information Module (Table 11-2). The PRIOR_INFORMATION_GROUPS input block (Table 11-3) contains information related to groups of prior-information equations. The LINEAR_PRIOR_INFORMATION input block (Table 11-4) contains information related to individual, linear prior-information equations.

Table 11-2. Input blocks read by the Prior-Information Module.

Input block	LLIST pointer	Read by subroutine	Required?[1]
PRIOR_INFORMATION_GROUPS	LLPTRGPPRIOR	PRI_INI_READ	No
LINEAR_PRIOR_INFORMATION	LLPTRPRIOR	PRI_INI_READ	No

[1] "No" means the REQUIRED argument in the call to UTL_READBLOCK is false; therefore, the input block is optional.

Chapter 11: Prior-Information Module

Table 11-3. Data of the PRIOR_INFORMATION_GROUPS input block.
[No default column order is defined for the PRIOR_INFORMATION_GROUPS input block. Table entries include all data recognized by the Prior-Information Module. "len" is the maximum number of characters in the string.]

Keyword or column name	Data type	Default	Description
GROUPNAME[1]	Character (len=12)	DefaultPrior	Group name.
USEFLAG	Logical	True	True means include group in analysis.
PLOTSYMBOL	Integer	1	Integer to be associated with group and group members in some data-exchange files (Chapter 17).
WTMULTIPLIER	Double precision	1.0	Value to be used to multiply weight of each group member.
COVMATRIX	Character (len=12)	No default	Name of variance-covariance matrix for group.

[1] GROUPNAME is the keyitem (Chapter 3) for the PRIOR_INFORMATION_GROUPS input block.

Table 11-4. Data of the LINEAR_PRIOR_INFORMATION input block.
[Column names are in default column order (Chapter 3) and include all data recognized by the Prior-Information Module. "len" is the maximum number of characters in the string.]

Keyword or column name	Data type	Default	Description
PRIORNAME[1]	Character (len=LENDNAM[2])	No default	Name of prior-information equation.
EQUATION	Character (len=MAX_STRING_LEN[2])	No default	The equation used to define how the parameters relate to the estimate of the prior information.
PRIORINFOVALUE	Double precision	0.0	Estimated value of prior-information equation.
STATISTIC	Double precision	No default	Statistic from which a weight is to be calculated (Table 10-1).
STATFLAG	Character (len=6)	No default	Flag indicating what STATISTIC is (Table 10-1).
GROUPNAME	Character (len=12)	"DefaultPrior"	Group name.

[1] PRIORNAME is the keyitem (Chapter 3) for the LINEAR_PRIOR_INFORMATION input block.
[2] See Chapter 5 and Appendix C.

Chapter 12: Parallel-Processing Module
By Edward R. Banta

Most applications constructed using the JUPITER API initiate execution of model(s) many times; in some situations, sets of model runs may be independent in that they can be executed concurrently. The Parallel-Processing Module can be used to advantage whenever dependent values from independent model runs are needed. This situation applies, for example, when sensitivities are calculated using perturbation methods or when Monte Carlo methods are used to generate many parameter sets for model input.

The Parallel-Processing Module contains data and subprograms to support parallel processing. The syntax to use the module is "USE PARALLEL_PROCESSING". Data and subprograms of the module are documented in Appendix J. The source code for the module is in a file named pll.f90. The Parallel-Processing Module uses the Datatypes, Global Data, Utilities, and Model Input-Output Modules, which need to be available when pll.f90 is compiled.

Program Design

Programs that require numerous runs of a model potentially can finish in less time if the model runs concurrently on multiple computers or multiple processors on one computer. The Parallel-Processing Module can be used to coordinate multiple computers and processors to make concurrent model runs. To make effective use of the parallel-processing capability provided by the Parallel-Processing Module, a model must be written in such a way that the data used to prepare model input and the data extracted from model output are independent in each model run. When multiple model runs are to be made in parallel, a single call is made to subroutine PLL_MAKE_RUNS. Many of the arguments required by PLL_MAKE_RUNS are likely to be available in any application that uses the JUPITER API for preparing model-input files, making model runs, and extracting model-calculated values from model-output files; complete documentation of PLL_MAKE_RUNS is provided in Appendix J. However, the following requirements of PLL_MAKE_RUNS are of particular importance to the issue of parallelization:
1. A full set of parameter values required for generating model-input files needs to be provided for each model run. The parameter values are passed in the argument list in a 2-dimensional array containing all parameter values required for each model run.
2. Accommodation for storage of the model-calculated values of interest for each model run needs to be provided. The model-calculated values are returned in the argument list in a 2-dimensional array, which is populated by the subroutine.

Method

The parallel-processing method supported by the current version of the JUPITER API is a technologically simple one that has advantages and disadvantages. Parallelization is implemented by executing a "dispatcher" program (for example, the SENSITIVITY_EXAMPLE application, Chapter 18) and one or more instances of a "runner" program (JRUNNER, Chapter 18). The dispatcher program can be considered to be the main program; the runner program is subordinate in that it performs duties as requested by the dispatcher. JRUNNER performs three tasks associated with a run of the process model as outlined in Table 2-3. The tasks are (1) adapt parameter values, (2) execute process model(s), and (3) extract values from model output.

For each processor or computer to be employed in parallel processing, JRUNNER is run in a unique directory. The directories in which JRUNNER is run may be on one computer or on multiple computers in a network. The dispatcher program needs to be able to write and read files in all the directories in which JRUNNER runs. When program input indicates that the parallel-processing option is to be used, at least one instance of JRUNNER needs to be running when execution of the dispatcher program is initiated. Additional instances of JRUNNER may be started after the dispatcher program is started.

Communication between the dispatcher and runner programs is implemented by creating, writing, reading, and deleting "signal" files, which are specially named files, some of which contain information to be communicated between the programs. Table 12-1 lists all signal files with their purposes. Note that a potential conflict exists if a runner directory contains a file with the same name as a signal file, if that file is needed by a program other than the dispatcher program or JRUNNER. When JRUNNER starts, it writes a signal file (named "jrunner.rdy") to indicate its active status to the dispatcher program, then it begins looking in its directory for signal files written by the dispatcher program. The dispatcher program executes the overall logic of the program, as is the case when the parallel-processing capability is not used. The dispatcher program writes a signal file named "jdispatch.rdy" in each runner directory to communicate initialization information, which is then read by JRUNNER. At a point in the dispatcher program where a set of model runs is required and can be made in parallel, subroutine PLL_MAKE_RUNS is called. PLL_MAKE_RUNS writes a signal file named "jdispar.rdy" to the directory of each active runner. Each copy of jdispar.rdy contains a set of parameter values and other data needed for a single model run. When JRUNNER detects the existence of a jdispar.rdy file in its directory, it reads the parameter values and other data from the file. JRUNNER uses the parameter values and template files to prepare one or more model-input files, then it sends a command to the operating system to initiate a model run. When the model run is complete, JRUNNER extracts model-calculated values from one or more model-output files and writes the values to a signal file named "jrundep.rdy". The dispatcher program iteratively checks the directory of each of the active runners for the presence of a file named "jrundep.rdy"; when it finds this file, it reads the model-calculated values from the file and populates a 1-dimensional section of an array with model-calculated values.

At this point, JRUNNER again begins to look for signal files written by the dispatcher program and to make model runs as needed. It continues this cycle until a file named "jdispatch.fin" is found; when this file is found, JRUNNER reads a single word from the file. JRUNNER expects the word to be either "STOP" or "RESET". If the word is "STOP", JRUNNER stops execution. If the word is "RESET", JRUNNER deallocates memory and starts looking for the "jdispatch.rdy" file, in preparation for a repetition of the initialization task. Note that the "jdispatch.fin" file can be prepared by any text editor, by a user-written utility program, or by an operating-system command such as "echo STOP > jdispatch.fin." When all the required model runs have been made and the 2-dimensional array of model-calculated values is fully populated, PLL_MAKE_RUNS returns control to the calling program unit.

In addition to the signal files written by the dispatcher program, JRUNNER requires one or more template file(s) for preparing model-input files (Chapter 8). Instructions for extracting model-calculated dependent values are transmitted in the "jdispatch.rdy" file, so instruction files are not required as input to JRUNNER. However, instruction files are required by the dispatcher program.

Table 12-1. Signal files for parallel processing.

File name	Variable used to store file name	Program that writes file	Purpose
jrunner.rdy	FNRUNRDY	JRUNNER	Tells dispatcher that runner is active and ready to make a model run.
jdispatch rdy	FNDISRDY	Dispatcher	Tells runner data needed to start model runs.
jdispar rdy	FNDISPAR	Dispatcher	Tells runner the command to be used to start a model run, parameter values to be used, and that runner should initiate a model run.
jrundep.rdy	FNRUNDEP	JRUNNER	Tells dispatcher model-calculated dependent values.
jdispatch fin	FNDISFIN	Dispatcher	Tells runner to stop execution.
jrunner.fin	FNRUNFIN	JRUNNER	Tells dispatcher that the signal to stop execution has been received, and that runner is stopping.
jrunfail fin	FNRUNFAIL	JRUNNER	Tells dispatcher that runner has encountered an error and is stopping.

All instances of JRUNNER prepare model-input files and extract model-calculated values from model-output files in the same way. For this approach to work correctly, when model runs are to be made in parallel, each runner directory needs to be set up so that the following requirements are met:

1. Generally, the runner directories and all directories containing files used by JRUNNER and the model need to be set up to be independent of each other, to avoid file-access conflicts as JRUNNER reads and writes files, and as the model reads and writes files. If the model is known to open input files in such a way that multiple instances of the model can access the same input file simultaneously without conflict, this restriction can be relaxed with respect to model-input files other than those to be created by JRUNNER.
2. A single operating-system command, which is provided by the dispatcher program in the "jdispatch.rdy" file and is the same for all runners, when issued from JRUNNER to the operating system, will start a model run.
3. Template files used in the construction of model-input files need to be available using a single set of file names provided in the TFILE argument of PLL_MAKE_RUNS. The names of the template, model-input, and model-output files are communicated to JRUNNER in the "jdispatch.rdy" file, so the file names (or relative pathnames) need to be valid in all the runner directories.
4. A single set of file names (or relative pathnames) provided in the MIFILE argument of PLL_MAKE_RUNS will generate files that the model will use as input files.
5. Model-output files need to be available (after being written by the model) using a single set of file names (or relative pathnames) provided in the MOFILE argument of PLL_MAKE_RUNS.
6. A single set of instructions, passed to PLL_MAKE_RUNS in the CINSTSET argument, for extracting model-calculated values from model-output files needs to be applicable in all runner directories. The instruction set is passed to JRUNNER in the jdispatch.rdy file, so instruction files are not required to be present in each runner directory.
7. The runner directories need to be accessible, from the directory where the dispatcher program runs, on a computer or within a network using pathnames, which may be absolute or relative. In computing environments that support it, the Universal Naming Convention (UNC) may be used to construct pathnames. (see http://www.webopedia.com/TERM/U/UNC.html)

8. For assured results when using the current version of the Parallel-Processing Module, the dispatcher program and all instances of JRUNNER need to be run under a common operating system. However, it may be possible to obtain successful results using a combination of closely related operating systems, such as different versions of Unix, or a combination of Unix and Linux. Experimentation is required to determine if a particular combination of operating systems can be successfully employed in parallel processing. The critical issue that determines if two operating systems can be used in combination is that text files written by the dispatcher program need to be readable by all of the runner programs, and vice versa. This restriction, for example, likely would prevent an arrangement where a dispatcher program is run on a Windows computer in combination with instances of JRUNNER running on Unix computers, due to the difference in the line-ending conventions used by the two systems.

Opening, writing, reading, and closing of disk files are processes that require a finite amount of time. Because the current parallel-processing capability of the Parallel-Processing Module communicates through disk files, a certain amount of delay time is built into the sections of code that deal with these files. The WAIT variable of the Parallel_Runners input block provides the user with a way to control the increment of time used in these sections of code. Generally, a small fraction of a second provides a sufficiently large time increment for the necessary file-handling delays—the default value of WAIT is 0.001 second. However, in some cases WAIT may need to be set to a larger value. The need to increase WAIT can be identified by the appearance of the following message on the screen running either the dispatcher program or JRUNNER: "Warning: WAIT time may be too small." Little is to be gained by setting WAIT smaller than the default value.

When JRUNNER encounters an error that it is designed to handle, it communicates an error message to the dispatcher program in the jrunfail.fin file and stops execution. When this situation arises, PLL_MAKE_RUNS reads the error message from the jrunfail.fin file and writes it to the screen.

Verbosity of messages and errors reported by JRUNNER is controlled by the VERBOSERUNNER variable, which is communicated by the dispatcher program to JRUNNER in the jdispatch.rdy file. If a problem is encountered by a user in implementing parallel processing in a particular situation, messages written by JRUNNER can be helpful in determining the cause of the problem. Setting VERBOSERUNNER to 5 will cause JRUNNER to write all optional messages.

Sometimes, JRUNNER will encounter an error that it is not designed to recognize, or it may stop due to an external problem such as a power interruption or network failure. In situations like these, no signal file is generated by JRUNNER to communicate to the dispatcher program that there is a problem. The Parallel-Processing Module is designed to handle situations like these by comparing the elapsed time since writing the jdispar.rdy file to an expected run time. If the elapsed time exceeds the expected run time by a user-defined factor (variable TIMEOUTFAC, default is a factor of 3), the model run is assumed to have failed, and that instance of JRUNNER is assumed to be inactive. The model run that failed is then assigned to the next available, active runner. The RUNTIME variable of the Parallel_Runners input block allows the user to specify an initial expected run time. As model runs are completed by each active runner, the expected run time is adjusted, so that the expected run time will approach the actual run time on each runner. Because the expected run times are continually being adjusted and a run is not assumed to have failed until the elapsed time exceeds TIMEOUTFAC times the expected run time, RUNTIME does not need to be a particularly accurate estimate of the model run time.

Chapter 12: Parallel-Processing Module

The method used by the parallel-processing capability implemented in the current version of the Parallel-Processing Module has both positive and negative attributes. It is straightforward to set up parallel processing because no additional software is needed. The method is fairly robust in that failure of individual runners can be tolerated, failed runners can be restarted without restarting the dispatcher program, and runners can be started after the dispatcher program has started. A major drawback of the method is that both the dispatcher program and JRUNNER use the central processing unit (CPU) of the computer virtually constantly while they are running, because they are at all times either running the model or they are testing for the existence of signal files. This attribute makes the use of the parallel-processing capability unwelcome in a computing environment shared among multiple users. However, in environments where exclusive use of multiple computers is acceptable, the savings in execution times can be substantial. Some operating systems provide a utility that can be used to set the priority of processes running on the computer. For example, on some Windows operating systems, the Task Manager utility provides this capability. In some cases, by setting the priority of the dispatcher and runner programs to have a priority level below the default level, the problem of these processes monopolizing the CPU can be reduced to an acceptable level.

In the simplest case, instances of JRUNNER are started manually by typing "JRUNNER" at the command prompt or by invoking JRUNNER from a script or batch file. Alternatively, utilities for starting a program remotely could be used to make the starting of JRUNNER instances more convenient. However, the issue of starting programs remotely is beyond the scope of the JUPITER API.

Public Subprograms

The public subprograms of the Parallel-Processing Module are listed in Table 12-2; detailed descriptions of the subprograms are in Appendix J.

Table 12-2. Public subprograms of the Parallel-Processing Module.
[Subprograms are listed in the order in which they typically would be called. Subprogram naming conventions are described in Table 2-3.]

Subprogram name	Purpose
Subprograms to be called by dispatcher program	
PLL_INI_DISPATCHER	Read input blocks labeled "PARALLEL_CONTROL" and "PARALLEL_RUNNERS", allocate and populate module arrays, determine status of each runner, and create "jdispatch rdy" files.
PLL_MAKE_RUNS	Create signal files to instruct runners to make model runs and read signal files created by runners; requires previous call to PLL_INI_DISPATCHER.
PLL_STOP_RUNNERS	Write "jdispatch.fin" file in directories of all active runners to instruct runner programs to stop execution.
Subprograms to be called by runner program	
PLL_INI_RUNNER_DIM	Look for "jdispatch.rdy" file; when found, read dimensioning variables from it and initialize scalar variables of the Model Input-Output Module.
PLL_INI_RUNNER_POP	Read array values from "jdispatch rdy" file and populate arrays, then close and delete the file; initialize and populate arrays of the Model Input-Output Module; requires previous call to PLL_INI_RUNNER_DIM.
PLL_READ_DISPAR	Read and store the contents of a "jdispar.rdy" file, then delete the file.
PLL_ADA	Write model-input files; requires previous calls to MIO_INI_ALLOC, MIO_INI_INPUTFILES_RUNNER, and MIO_INI_TEMPLATE.
PLL_EXE	Submit command to start a model run to the operating system; requires previous call to PLL_READ_DISPAR.
PLL_EXT	Extract model-calculated values from model-output files; requires previous calls to MIO_INI_OUTPUTFILES_RUNNER and MIO_INI_INSTRUCT1.
PLL_WRITE_RUNDEP	Write model-calculated values to "jrundep.rdy" file.
PLL_RUNNER_STOP	Delete unneeded signal files and stop execution.
Subprograms to be called by either dispatcher or runner program	
PLL_WAIT	Execute a loop continuously until a specified amount of time has elapsed.
PLL_CLN	Deallocate all arrays of the module.

Input Blocks Read by the Parallel-Processing Module

Two input blocks may be read by subroutine PLL_INI_DISPATCHER (Table 12-3). The PARALLEL_CONTROL input block (Table 12-4) is intended to contain information related to the operation of the parallel-processing capability provided by the JUPITER API. The PARALLEL_RUNNERS input block (Table 12-5) is intended to contain information related to the computing environment of individual runners; both are optional. After PLL_INI_DISPATCHER has been called and these input blocks have been read, data read from the PARALLEL_CONTROL block are accessible using the pointer LLPTRPLLCTRL, and data read from the PARALLEL_RUNNERS block are accessible using the pointer LLPTRPLLRUN.

Table 12-3. Input blocks read by the Parallel-Processing Module.

Input block	LLIST pointer	Read by subroutine	Required?[1]
PARALLEL_CONTROL	LLPTRPLLCTRL	PLL_INI_DISPATCHER	No
PARALLEL_RUNNERS	LLPTRPLLRUN	PLL_INI_DISPATCHER	No

[1] "No" means the REQUIRED argument in the call to UTL_READBLOCK is false; therefore, the input block is optional.

Table 12-4. Data of the PARALLEL_CONTROL input block.
[No default column order is defined for the PARALLEL_CONTROL input block. Table entries include all data recognized by the Parallel-Processing Module. "len" is the maximum number of characters in the string.]

Keyword or column name	Data type	Default	Description
PARALLEL	Logical	False	"True" activates parallel processing.
OPERATINGSYSTEM	Character (len=20)	"WINDOWS"	Operating system under which the dispatcher and runner programs run. Valid values are "WINDOWS", "DOS", "UNIX", and "LINUX".
WAIT	Floating-point	0.001	Time delay increment (seconds) used as needed in checking for presence of, and reading from, signal files.
VERBOSERUNNER	Integer	3	"Verbose" setting (IVERB of the Global Data Module, Chapter 5 and Appendix C) for runners.
AUTOSTOPRUNNERS	Logical	True	"True" causes subroutine PLL_STOP_RUNNERS (when called) to send signal to runners to stop execution. "False" causes PLL_STOP_RUNNERS to send signal to runners to reset and continue execution.
TIMEOUTFACTOR	Floating-point	3.0	Factor that multiplies RUNTIME to determine if a model run is overdue; assigned to module variable TIMEOUTFAC.

Table 12-5. Data of the PARALLEL_RUNNERS input block.
[Column names are in default column order (Chapter 3) and include all data recognized by the Parallel-Processing Module. "len" is the maximum number of characters in the string.]

Keyword or column name	Data type	Default	Description
RUNNERNAME[1]	Character (len=max_string_len[2])	no default	Name by which runner will be identified.
RUNNERDIR	Character (len=max_string_len[2])	no default	Name (can be a pathname) of directory where runner program runs; name needs to end in directory-separator character used by the operating system, either "\" or "/".
RUNTIME	Double precision	10	Expected model run time (seconds).
RENAME	Character (len=6)	"ren"	Operating-system command to rename a file.

[1]RUNNERNAME is the keyitem (Chapter 3) for the PARALLEL_RUNNERS input block.
[2]See Chapter 5 and Appendix C.

Chapter 13: Sensitivity Module
By Edward R. Banta and Eileen P. Poeter

The purpose of the Sensitivity Module is to populate a sensitivity matrix (a double-precision 2-D array) with sensitivities of simulated equivalents to observations and (or) predictions with respect to parameters. The Sensitivity Module is designed to be used in the context of an application program structured as described in Chapter 2 and exemplified in the "Example Application: SENSITIVITY_EXAMPLE" section of Chapter 18. It provides subroutines and variables needed to populate the sensitivity matrix. The sensitivities can be calculated by perturbation, extracted from model-output files, or both. The Sensitivity Module contains no public data. Data and subprograms of the Sensitivity Module are documented in Appendix K. The source code for the Sensitivity Module is in a file named sen.f90. The Sensitivity Module uses the Datatypes, Global Data, Utilities, Basic, and Dependents Modules, which need to be available when sen.f90 is compiled.

When sensitivities are to be calculated by perturbation of parameter values, subroutines of the Sensitivity Module are used to define the job of the Control Loop, to generate sets of parameter values with each set containing one perturbed parameter value, to select the appropriate model command, and to populate the sensitivity matrix. In this case, the sensitivities are calculated by finite differences based on the unperturbed and perturbed parameter values and the model-calculated (dependent) values from model runs using the parameter-value sets. The Sensitivity Module supports calculation of sensitivities by both the forward-difference and the central-difference methods, as described in Table 13-1. Calculation of sensitivities by the central-difference method requires an additional model run for each parameter for which perturbation sensitivities are required and thus is more time-consuming than calculation of sensitivities by the forward-difference method. However, sensitivities calculated by the central-difference method are in general more accurate than those calculated by the forward-difference method.

Table 13-1. Methods of calculating perturbation sensitivities.

Method of calculating perturbation sensitivities	Formula[1]
Forward-difference method	$\left(D_f - D_o\right)/\left(B_f - B_o\right)$
Central-difference (2-point) method	$\left(D_f - D_b\right)/\left(B_f - B_b\right)$

[1] B_o is the unperturbed value of parameter B, B_f is the forward-perturbed parameter value (B_o+INC), B_b is the backward-perturbed parameter value (B_o–INC), INC is an incremental change to the parameter value, D_o is the value of dependent D generated by running the model using unperturbed parameter values, D_f is the value of dependent D generated by running the model using parameter value B_f substituted for B_o, and D_b is the value of dependent D generated by running the model using parameter value B_b substituted for B_o.

When sensitivities are to be calculated by the process model, subroutines of the Sensitivity Module are used to define the job of the Control Loop, to select the appropriate model command, and to populate the sensitivity matrix with model-calculated sensitivities. In this case, the Sensitivity Module reads the derivatives interface file named in the DERIV_INTERFACE variable of the Basic Module (Chapter 7 and Appendix E), opens the file containing model-calculated sensitivities, reads the sensitivities from that file, and uses those sensitivities to populate the sensitivity matrix.

Chapter 13: Sensitivity Module

In some circumstances it may be desirable to obtain sensitivities for some parameters from the model and to calculate sensitivities for other parameters by perturbation. When this situation arises, the Sensitivity Module enables the application to use both methods to populate the sensitivity matrix. The SENSITIVITY_EXAMPLE program described in Chapter 18 illustrates the use of the Sensitivity Module to support the use of perturbation, model calculation of sensitivities, or a combination of both methods to populate the sensitivity matrix.

Public Subprograms

The public subprograms of the Sensitivity Module are listed in Table 13-2; detailed descriptions of the subprograms are in Appendix K.

Table 13-2. Public subprograms of the Sensitivity Module.
[Subprograms are listed in order in which they typically would be called. Subprogram naming conventions are described in Table 2-3]

Subprogram name	Purpose
SEN_INI	Allocate module arrays and read a derivatives interface file.
SEN_DEF	Define job of Control Loop when sensitivities are to be calculated; requires previous call to SEN_INI.
SEN_GEN	Generate a set of parameter values, optionally with perturbation of one parameter; if a parameter is perturbed, store the perturbation amount; requires previous call to SEN_DEF.
SEN_EXE_SELECT	Assign number of command to be executed for calculation of sensitivities; requires previous calls to BAS_INI_MODELEXEC and SEN_DEF.
SEN_UEV_POPX_MODCALC	Populate rows of a sensitivity array with model-calculated derivatives, requires previous call to SEN_INI.
SEN_UEV_POPXROW_DIFF	Populate one row (for one parameter) of a sensitivity array by finite differences; requires previous call to SEN_GEN.
SEN_UEV_LNX	Convert native-space derivatives to log-space derivatives for specified parameters.
SEN_UEV_WRITESENTABLE	Write a table of dependent-variable sensitivities.
SEN_UEV_DX_WRITE_SEN	Write "_s1", "_sc", "_sd", "_so", and "_su" data-exchange files (Chapter 17); requires previous call to DEP_INI_READ.
SEN_UEV_DX_WRITE_MATRIX	Write either "_s1", "_sd", or "_su" data-exhange file (Chapter 17) open on specified file unit.
SEN_UEV_DX_READ_SU	Read "_su" data-exchange file (Chapter 17).
SEN_UEV_DX_READ_MATRIX	Read either "_s1", "_sd", or "_su" data-exhange file (Chapter 17) open on specified file unit.
SEN_CLN	Deallocate all module allocatable arrays.

Chapter 14: Statistics Module
By Eileen P. Poeter and Mary C. Hill

The Statistics Module calculates and writes statistics related to fit of simulated equivalents to observations and of parameter values to prior information. The Statistics Module contains a small amount of public data. Data and subprograms of the Statistics Module are documented in Appendix L. The source code for the Statistics Module is in a file named sta.f90. The Statistics Module uses the Datatypes, Global Data, Utilities, Basic, and Dependents Modules, which need to be available when sta.f90 is compiled. The statistics included in this module are calculated as described by Hill and Tiedeman (in press) and listed in Table 14-1.

Table 14-1. Statistics calculated or used by the Statistics Module.
[More information on these statistics can be found in many texts, including Hill and Tiedeman (in press) and Poeter and others (in press).]

Statistic	Reference, Expression, or Explanation
95 percent confidence interval on parameters	1.96 standard deviations above and below the optimal parameters value.
SSWR	Sum of weighted squared residuals. Values for observations only, prior information only, and the total value are calculated.
Calculated error variance	$$\frac{SSWR}{NOBS + MPR - NPERD}$$ Where, $NOBS$ = number of observations, MPR = number of prior-information equations, and $NPERD$ =number of estimated parameters
Standard error of the regression	$(\text{Calculated error variance})^{0.5}$
MLOF (dependents)	Maximum Likelihood Objective Function. Calculated as $$NOBS \ln(2\pi) + \ln|\omega| + \frac{SSWR}{\sigma}$$ where, ω = weight matrix, including division by any defined common error variance; $NOBS$ = (number of observations used)
MLOF (dependents and prior)	As above, but replacing $NOBS$ with n. n = (number of observations used) + (number of prior-information equations used)
AICc	Corrected Akaike's Information Criterion. Calculated as $$MLOF + 2k + \left(\frac{2k(k+1)}{NOBS - k - 1}\right)$$ where $MLOF$ is for dependents only and k = NRERD+1
BIC	Bayesian Information Criterion. Calculated as $MLOF + k \log(NOBS)$ where $MLOF$ is for dependents only
HQ	Hannan-Quin Information Criterion. Calculated as $MLOF + 2k \log(\log(NOBS))$ where $MLOF$ is for dependents only

Statistic	Reference, Expression, or Explanation
KIC	Kashyap's Information Criterion. Calculated as $$MLOF + k \log\left(\frac{NOBS}{2\pi}\right) + \mid X^T \omega X \mid$$ where $MLOF$ is for dependents only; X =sensitivity matrix, and X^T is its transpose
Log determinant of Fisher information matrix	$\log_e \mid X^T \omega X \mid$
R2N (dependents)	Correlation of weighted residuals for observations to standard normal statistics.
R2N (dependents and prior)	Correlation of weighted residuals for observations and prior-information equations to standard normal statistics.
Runs statistic	Indicates the probability that the number of runs reflect that the values are randomly distributed.

Public Subprograms

The public subprograms of the Statistics Module are listed in Table 14-2; detailed descriptions of the subprograms are in Appendix L.

Table 14-2. Public subprograms of the Statistics Module.
[Subprograms are listed in the order in which they typically would be called. Subprogram naming conventions are described in Table 2-3]

Subprogram name	Purpose
STA_INI	Allocate memory and calculate weighted observed values.
STA_UEV_INIT	Initialize variables for the Use Extracted Values task of the module.
STA_UEV_FIT	Write simulated and observed values, and calculate and write statistics related to residuals; requires previous calls to STA_INI and DEP_INI_READ.
STA_UEV_DX_WRITE_NM	Write "_nm" data-exchange file (Chapter 17); requires previous call to DEP_INI_READ.
STA_UEV_DX_WRITE_DM	Write "_dm" data-exchange file (Chapter 17).
STA_UEV_DX_READ_DM	Read "_dm" data-exchange file (Chapter 17).
STA_EVA_RUNS	Calculate and write the runs statistic for all residuals.
STA_EVA_ORDER	Order residuals and corresponding names, and populate an array with standard normal statistics; requires previous call to DEP_INI_READ.
STA_EVA_PROB_NORM_DISTRB	Find the probability related to a specified U value, or the U value related to a specified probability for a standard Gaussian distribution, where U is a number of standard deviations.
STA_EVA_COMMENTS	Write comments on interpretation of the R2N statistic.
STA_EVA_CRITVAL_R2N	Calculate critical value of the R2N statistic.
STA_CLN	Deallocate the module array.

Chapter 15: Parameter-Value Generation
By Mary C. Hill, Edward R. Banta, Eileen P. Poeter, and John E. Doherty

Applications developed using the JUPITER API are likely to require functionality that may be categorized in the "Generate Parameter Values" task, which is conceptually described in this chapter. The purpose of the Generate Parameter Values task is to generate one or more sets of parameter values as needed to fulfill the goals of the application. For convenience of discussion in this chapter, it is assumed that the application would include a Fortran module for each approach to parameter-value generation required by the application; however, use of the JUPITER API is not restricted to this approach. It is anticipated that research efforts involving the use of the JUPITER API commonly would include programming of new modules or modification of existing modules in this task. Modules that could be assigned to this task include but are certainly not limited to:

- Modified Gauss Newton, as in MODFLOW-2000 (Hill and others, 2000) and UCODE_2005 (Poeter and others, 2005);
- Modified Gauss Newton, as in PEST (Doherty, 2004);
- Trust region version of Modified Gauss Newton as in UCODE_2005 (Poeter and others, 2005);
- Markov Chain Monte Carlo; and
- Other Monte Carlo methods of generating parameter values.

An application might use all of these modules, a subset of them, or none of them; the purpose of an application is largely determined by the parameter-value generator it uses.

Specific parameter generators require information about each parameter, which can be specific to each module. For example, in defining parameters UCODE_2005 uses SCALEPVAL, SENMETHOD, TRANSFORM, and TOLPAR; whereas, PEST uses PARTRANS, PARCHGLIM, PARGP, SCALE, OFFSET, and DERCOM. In general, other modules would not need access to these additional pieces of information. That being the case, each parameter-generator module likely would allocate memory, read, and store that additional information in arrays of that particular module. Generally, only that module would need to have access to that information. However, it may be advantageous to users to be able to list data for multiple modules in a single input file. The use of input blocks for entering parameter-related information would facilitate the sharing of this information among multiple modules, by having each module extract the information it needs from an input-block data structure containing parameter information.

Chapter 16: Input Block Names and Keywords
By John E. Doherty, Eileen P. Poeter, and Edward R. Banta

Input Block Names and Ordering

A particular application that uses the JUPITER API can have whatever input blocks are desired by the developer. However, the adoption of a uniform set of input-block names and an order for those input blocks will facilitate the use of input files by multiple applications. The order of input blocks as listed in Table 16-1 is the conventional order where the input blocks named are to be read from a single input file. However, there is no requirement that an application read all input blocks from a single input file. The order in which input blocks are listed in a particular input file will dictate the order in which the input blocks will be read by calls to UTL_READBLOCK (Chapter 6 and Appendix D) into input-block data structures associated with pointers of type LLIST (Chapters 3 and 4). However, the data stored in this form can be accessed using calls to UTL_FILTER and UTL_FILTERLIST (Chapter 6 and Appendix D) in any order. The order of input blocks in Table 16-1 was developed with the understanding that the required presence of input blocks listed later may depend on the contents of input blocks listed earlier. The dependence of the order on requirements related to memory allocation or other considerations is considered secondary, because access to the data (using UTL_FILTER and UTL_FILTERLIST) is independent of the order in which the input blocks are read from an input file. Not all input blocks listed in Table 16-1 will be required by all applications, and not all of them are read or used by the JUPITER API as defined in this document.

79

Table 16-1. Blocklabels for input blocks defined in the JUPITER API.
[The order of input blocks in this table is the conventional order for applications that use the JUPITER API. Grey shading identifies input blocks needed for most applications (as qualified by footnotes 6 and 7). The other input blocks are optional]

Blocklabel	Purpose	Default columns[1]	Module (Chapter)
Options[2,3]	Define application operation	No	Basic (7)
App_Control_Data[2,3,4] (for example; blocklabel would be application-defined)		Application-defined	
App_Controls[3] (for example; blocklabel would be application-defined)		Application-defined	
Model_Command_Lines[2,3]		No	Basic (7)
Parameter_Groups[2,3]	Define parameters	Application-defined (see "Example Application: SENSITIVITY_EXAMPLE" section of Chapter 18)	
Parameter_Data[2,3]			
Parameter_Values[2,3]			
Derived_Parameters[3]			
Observation_Groups[3,5]	Define observations	No	Dependents (10)
Observation_Data[2,3,5,6]		Yes	
Derived_Observations[3]		Yes	
Prediction_Groups[3]	Define predictions	No	Dependents (10)
Prediction _Data[3,7]		Yes	
Derived_ Predictions[3]		Yes	
Prior_Information_Groups[3]	Define prior information	No	Prior Information (11)
Linear_Prior_Information[3]		Yes	
Matrix_Files[2,3]	Define matrices.[8]	No	Basic (7)
Model_Input_Files[2,3]	Interact with model input and output files	Yes	Model Input-Output (8)
Model_Output_Files[2,3]		Yes	
Parallel_Control[2,3]	Invoke multiple model runs concurently	No	Parallel Processing (12)
Parallel_Runners[2,3]		No	

[1]"Default columns" applies to input blocks for which blockformat is specified as TABLE (see "Blockformat" section, below) Yes: The input block has a default column order; these blocks do not

require columns to be labeled. No: The input block has no default column order; column labels are required when blockformat is specified as TABLE.
[2]See "Example Application: SENSITIVITY_EXAMPLE" section of Chapter 18 for an example of this type of input block.
[3]See Poeter and others (2005), UCODE_2005 manual for an example of this type of input block.
[4]Applications generally require input block(s) here to control operation.
[5]See "Example Application: GROUP_EXAMPLE" section of Chapter 18 for an example of this type of input block.
[6]Only required if observations are supported.
[7]Only required if predictions are supported.
[8]In the JUPITER API, matrices are used to weight groups of observations or prior information with correlated errors.

Input blocks and their associated data structures that are generated by subroutines included in the JUPITER API are documented in Tables 16-2 through 16-20. Each table documents one input block, showing the blocklabel for the input block, the module where the LLIST pointer is declared, the subroutine from which UTL_READBLOCK is called, the subroutine where keywords are recognized (if different from the subroutine from which UTL_READBLOCK is called), the keyitem (or "not defined" if there is no keyitem), the name of the LLIST pointer, recognized keywords, the array or variable where values associated with each keyword are stored, and the access to the array or variable. If the keyitem is defined, the variable associated with the KLISTS argument of UTL_READBLOCK also is shown; this variable is used to dimension the arrays listed in the table.

Some input blocks can be used to define groups, which can have members defined in another input block. Input blocks that define groups have "GROUPNAME" as the keyitem. These input blocks serve one or both of two purposes: (1) Default values for variables recognized in the members input block can be defined to apply to all members of each group; and (2) values that apply to a group as a whole are efficiently stored in arrays dimensioned with the number of groups. The first purpose always applies when two input blocks have a group/member relation. The second purpose applies when values are stored in one or more arrays dimensioned to hold one value for each group. The assignment of default attributes defined by group to members of each group is accomplished by a call to UTL_GROUPLIST and requires that the input block of members include the keyword or column name GROUPNAME. Matching GROUPNAME values are used to relate members to groups; the comparison is case-insensitive. In tables for "group" input blocks (Tables 16-4, 16-8, 16-11, and 16-15), the associated "members" input block is identified by an explanatory entry at the bottom of the keyword column.

Table 16-2. "OPTIONS" input block.

Input block: **OPTIONS**
Module: **Basic**
Subroutine where UTL_READBLOCK is called: **BAS_INI_GETOPTIONS**
Keyitem: **Not defined**
Default column order: **Not defined**
Pointer to LLIST structure: **LLPTROPT**

Keyword	Variable used to store associated value	Access to variable
VERBOSE	IVERB of the Global Data Module	PUBLIC
DERIVATIVES_INTERFACE	DERIV_INTERFACE	PUBLIC

Table 16-3. "MODEL_COMMAND_LINES" input block.

Input block: **MODEL_COMMAND_LINES**
Module: **Basic**
Subroutine where UTL_READBLOCK is called: **BAS_INI_MODELEXEC**
Keyitem: **COMMAND**
Default column order: **COMMAND, PURPOSE, COMMANDID**
Array dimension (KLISTS argument): **NCOMLINES**
Pointer to LLIST structure: **LLPTRMODCOM**

Keyword	Array used to store associated value	Access to array
COMMAND	MODCOMLINE	PRIVATE
PURPOSE	COMPURPOSE	PUBLIC
COMMANDID	COMMANDID	PUBLIC

Table 16-4. "PARAMETER_GROUPS" input block.

Input block: **PARAMETER_GROUPS**
Module: **Application-defined**
Subroutine where UTL_READBLOCK is called: **Application-defined**
Keyitem: **GROUPNAME**
Default column order: **Application-defined**
Array dimension (KLISTS argument): **Application-defined**
Pointer to LLIST structure: **Application-defined**

Keyword	Array used to store associated value	Access to array
GROUPNAME	Application-defined	Application-defined
Any keyword recognized in "Parameter_Data" input block (Table 16-5), which contains group members	Application-defined	Application-defined

Table 16-5. "PARAMETER_DATA" input block.

Input block: **PARAMETER_DATA**
Module: **Application-defined**
Subroutine where UTL_READBLOCK is called: **Application-defined**
Keyitem: **PARAMNAME**
Default column order: **Application-defined**
Array dimension (KLISTS argument): **Application-defined**
Pointer to LLIST structure: **Application-defined**

Keyword	Array used to store associated value	Access to array
PARAMNAME	Application-defined	Application-defined
GROUPNAME	Application-defined	Application-defined
TRANSFORM	Application-defined	Application-defined
ADJUSTABLE	Application-defined	Application-defined
STARTVALUE	Application-defined	Application-defined
LOWERBOUND	Application-defined	Application-defined
UPPERBOUND	Application-defined	Application-defined

Table 16-6. "PARAMETER_VALUES" input block.

Input block: **PARAMETER_VALUES**		
Module: Application-defined		
Subroutine where UTL_READBLOCK is called: Application-defined		
Keyitem: PARAMNAME		
Default column order: Application-defined		
Array dimension (KLISTS argument): Application-defined		
Pointer to LLIST structure: Application-defined		
Keyword	**Array used to store associated value**	**Access to array**
PARAMNAME	Application-defined	Application-defined
STARTVALUE	Application-defined	Application-defined

Table 16-7. "DERIVED_PARAMETERS" input block.

Input block: **DERIVED_PARAMETERS**		
Module: Application-defined		
Subroutine where UTL_READBLOCK is called: Application-defined		
Keyitem: PARAMNAME		
Default column order: Application-defined		
Array dimension (KLISTS argument): Application-defined		
Pointer to LLIST structure: Application-defined		
Keyword	**Array used to store associated value**	**Access to array**
PARAMNAME	Application-defined	Application-defined
EQUATION	Application-defined	Application-defined

Table 16-8. "OBSERVATION_GROUPS" input block.

Input block: **OBSERVATION_GROUPS**		
Module: Dependents		
Subroutine where UTL_READBLOCK is called: DEP_INI_READ		
Keyitem: GROUPNAME		
Default column order: Not defined		
Array dimension (KLISTS argument): NGPS		
Pointer to LLIST structure: LLPTRDEPGP		
Keyword	**Array used to store associated value**	**Access to array**
GROUPNAME	GROUPNAM	PUBLIC
USEFLAG	USEFLAG	PRIVATE
PLOTSYMBOL	PLOTSYMBOL	PRIVATE
WTMULTIPLIER	WTMULTIPLIER	PRIVATE
MEASDIVIDER	MEASDIVIDER	PRIVATE
COVMATRIX	COVMATNAM	PRIVATE
Any keyword recognized in "Observation_Data" input block (Table 16-9), which contains group members	See Table 16-9	See Table 16-9

Table 16-9. "OBSERVATION_DATA" input block.

Input block: **OBSERVATION_DATA**
Module: **Dependents**
Subroutine where UTL_READBLOCK is called: **DEP_INI_READ**
Subroutine where keywords are recognized: **DEP_INI_STORE**
Keyitem: **OBSNAME**
Default column order: **OBSNAME, OBSVALUE, STATISTIC, STATFLAG, GROUPNAME, EQUATION**
Array dimension (KLISTS argument): **NTOTDEP**
Pointer to LLIST structure: **LLPTRDEP**

Keyword	Array used to store associated values	Access to array
OBSNAME	OBSNAM	DEP_INI_STORE argument list
OBSVALUE	OBSVAL	PUBLIC
STATISTIC	Used to populate VARIANCE array	PRIVATE
STATFLAG		
GROUPNAME	GROUPNOBS	PUBLIC
EQUATION	DERDEPEQN	PRIVATE
NONDETECT[1]	NONDETVAL	PUBLIC
WTOSCONSTANT[2]	WTCOS	PUBLIC

[1]If NONDETECT is found and is true, the value associated with OBSVALUE is stored in NONDETVAL.
[2]If WTOSCONSTANT is found and is not zero, the corresponding STATFLAG value must not be "CV".

Table 16-10. "DERIVED_OBSERVATIONS" input block.

Input block: **DERIVED _OBSERVATIONS**
Module: **Dependents**
Subroutine where UTL_READBLOCK is called: **DEP_INI_READ**
Subroutine where keywords are recognized: **DEP_INI_FILTERLIST**
Keyitem: **OBSNAME**
Default column order: **OBSNAME, OBSVALUE, STATISTIC, STATFLAG, GROUPNAME, EQUATION**
Array dimension (KLISTS argument): **NTOTDEP**
Pointer to LLIST structure: **LLPTRDEP**

Keyword	Array used to store associated values	Access to array
All keywords recognized in the Observation_Data input block (Table 16-9)	All entries are defined in Table 16-9	

Table 16-11. "PREDICTION_GROUPS" input block.

Input block: PREDICTION_GROUPS
Module: Dependents
Subroutine where UTL_READBLOCK is called: DEP_INI_READ
Keyitem: GROUPNAME
Default column order: Not defined
Array dimension (KLISTS argument): NGPS
Pointer to LLIST structure: LLPTRDEPGP

Keyword	Array used to store associated value	Access to array
GROUPNAME	GROUPNAM	PUBLIC
USEFLAG	USEFLAG	PRIVATE
PLOTSYMBOL	PLOTSYMBOL	PRIVATE
WTMULTIPLIER	WTMULTIPLIER	PRIVATE
MEASDIVIDER	MEASDIVIDER	PRIVATE
COVMATRIX	COVMATNAM	PRIVATE
Any keyword recognized in "Prediction_Data" input block, which contains group members	See Table 16-12	See Table 16-12

Table 16-12. "PREDICTION_DATA" input block.

Input block: PREDICTION_DATA
Module: Dependents
Subroutine where UTL_READBLOCK is called: DEP_INI_READ
Subroutine where keywords are recognized: DEP_INI_STORE
Keyitem: PREDNAME
Default column order: PREDNAME, REFVALUE, STATISTIC, STATFLAG, GROUPNAME, EQUATION
Array dimension (KLISTS argument): NTOTDEP
Pointer to LLIST structure: LLPTRDEP

Keyword	Array used to store associated values	Access to array
PREDNAME	OBSNAM	DEP_INI_STORE argument list
REFVALUE	OBSVAL	PUBLIC
STATISTIC	Used to populate VARIANCE array	PRIVATE
STATFLAG		
GROUPNAME	GROUP	PUBLIC
EQUATION	DERDEPEQN	PRIVATE
WTOSCONSTANT[1]	WTCOS	PUBLIC

[1]If WTOSCONSTANT is found and is not zero, the corresponding STATFLAG value must not be "CV".

Table 16-13. "DERIVED_PREDICTIONS" input block.

Input block: Derived _Observations Module: Dependents Subroutine where UTL_READBLOCK is called: DEP_INI_READ Subroutine where keywords are recognized: DEP_INI_FILTERLIST Keyitem: OBSNAME Default column order: PREDNAME, REFVALUE, STATISTIC, STATFLAG, GROUPNAME, EQUATION Array dimension (KLISTS argument): NTOTDEP Pointer to LLIST structure: LLPTRDEP		
Keyword	**Array used to store associated values**	**Access to array**
All keywords recognized in the Prediction_Data input block (Table 16-12)	All entries are defined in Table 16-12	

Table 16-14. "MATRIX_FILES" input block.

Input block: "Matrix_Files" Module: Basic Subroutine where UTL_READBLOCK is called: BAS_INI_COVMAT Keyitem: MATRIXFILE Default column order: Not defined Array dimension (KLISTS argument): KFILES Pointer to LLIST structure: LLPTRMATFIL		
Keyword	**Array used to store associated value**	**Access to array**
MATRIXFILE	Used to populate COVMATARR array	PUBLIC
NMATRICES		

Table 16-15. "PRIOR_INFORMATION_GROUPS" input block.

Input block: PRIOR_INFORMATION_GROUPS Module: Prior Information Subroutine where UTL_READBLOCK is called: PRI_INI_READ Keyitem: GROUPNAME Default column order: Not defined Array dimension (KLISTS argument): NPRIGPS Pointer to LLIST structure: LLPTRGPPRIOR		
Keyword	**Array used to store associated value**	**Access to array**
GROUPNAME	PRIGROUPNAM	PRIVATE
USEFLAG	PRIUSEFLAG	PRIVATE
PLOTSYMBOL	PLOTSYMBOLPRIGP	PRIVATE
WTMULTIPLIER	PRIWTMULTIPLIER	PRIVATE
COVMATRIX	PRICOVMATNAM	PRIVATE
Any keyword recognized in "Linear_Prior_Information" input block, which contains group members	See Table 16-16	See Table 16-16

Table 16-16. "LINEAR_PRIOR_INFORMATION" input block.

Input block: **LINEAR_PRIOR_INFORMATION**
Module: **Prior Information**
Subroutine where UTL_READBLOCK is called: **PRI_INI_READ**
Keyitem: **PRIORNAME**
Default column order: **PRIORNAME, EQUATION, PRIORINFOVALUE, STATISTIC, STATFLAG, GROUPNAME**
Array dimension (KLISTS argument): **MPR**
Pointer to LLIST structure: **LLPTRPRIOR**

Keyword	Array used to store associated value	Access to array
PRIORNAME	PRINAM	PUBLIC
EQUATION	PRIEQTEXT	PRIVATE
PRIORINFOVALUE	PRIVAL	PUBLIC
STATISTIC	Used to populated PRIVARIANCE array	PRIVATE
STATFLAG		
GROUPNAME	PRIGROUP	PRIVATE

Table 16-17. "MODEL_INPUT_FILES" input block.

Input block: **MODEL_INPUT_FILES**
Module: **Model Input-Output**
Subroutine where UTL_READBLOCK is called: **MIO_INI_INPUTFILES**
Subroutine where keywords are recognized: **MIO_INI_INPUTFILES**
Keyitem: **MODINFILE**
Default column order: **MODINFILE, TEMPLATEFILE**
Array dimension (KLISTS argument): **NUMINFILE**
Pointer to LLIST structure: **LLPTRMODIN**

Keyword	Array used to store associated value	Access to array
MODINFILE	MODINFILE	PRIVATE
TEMPLATEFILE	TEMPFILE	PRIVATE

Table 16-18. "MODEL_OUTPUT_FILES" input block.

Input block: **MODEL_OUTPUT_FILES**
Module: **Model Input-Output**
Subroutine where UTL_READBLOCK is called: **MIO_INI_OUTPUTFILES**
Subroutine where keywords are recognized: **MIO_INI_OUTPUTFILES**
Keyitem: **MODOUTFILE**
Default column order: **MODOUTFILE, INSTRUCTIONFILE, CATEGORY**
Array dimension (KLISTS argument): **NUMOUTFILE**
Pointer to LLIST structure: **LLPTRMODOUT**

Keyword	Array used to store associated value	Access to array
MODOUTFILE	MODOUTFILE	PRIVATE
INSTRUCTIONFILE	INSFILE	PRIVATE
CATEGORY	CATEGORY	PRIVATE

Table 16-19. "PARALLEL_CONTROL" input block.

Input block: **PARALLEL_CONTROL**
Module: **Parallel Processing**
Subroutine where UTL_READBLOCK is called: **PLL_INI_DISPATCHER**
Subroutine where keywords are recognized: **PLL_INI_DISPATCHER**
Keyitem: **Not defined**
Default column order: **Not defined**
Pointer to LLIST structure: **LLPTRPLLCTRL**

Keyword	Variable used to store associated value	Access to variable
PARALLEL	DOPLL	PRIVATE
WAIT	WAIT	PUBLIC
VERBOSERUNNER	IVERBRUNNER	PRIVATE
AUTOSTOPRUNNERS	AUTOSTOPRUNNERS	PUBLIC
OPERATINGSYSTEM	OS_PLL	PRIVATE
TIMEOUTFACTOR	TIMEOUTFAC	PRIVATE

Table 16-20. "PARALLEL_RUNNERS" input block.

Input block: **PARALLEL_RUNNERS**
Module: **Parallel Processing**
Subroutine where UTL_READBLOCK is called: **PLL_INI_DISPATCHER**
Subroutine where keywords are recognized: **PLL_INI_DISPATCHER**
Keyitem: **RUNNERNAME**
Default column order: **RUNNERNAME, RUNNERDIR, RUNTIME**
Array dimension (KLISTS argument): **NUMRUNNERS**
Pointer to LLIST structure: **LLPTRPLLRUN**

Keyword	Array used to store associated value	Access to array
RUNNERNAME	RUNNERNAME	PRIVATE
RUNNERDIR	RUNNERDIR	PRIVATE
RUNTIME	RUNTIME	PRIVATE

Keywords and Definitions

Each application would recognize a specified set of one or more keywords in each input block that it recognizes. Certain keywords are expected to have the same meaning in all applications where they are recognized, as indicated below. To facilitate interchangeability where possible, these keywords (Table 16-21), rather than application-specific keywords that mean the same thing, should be used.

Table 16-21. Keyword definitions.
["len" is the maximum number of characters in the string.]

Keyword or Column name	Data type	Definition and, if applicable, valid values
ADJUSTABLE	Logical[1]	If true, parameter value may be adjusted by application during analysis.
AUTOSTOPRUNNERS	Logical[1]	If true, runners (instances of JRUNNER program, Chapter 12) are notified to stop execution when PLL_STOP_RUNNERS (Appendix J) is called. If false, runners are notified to reset themselves.
CATEGORY	Character(len=6)	Identifies type of dependent, either "obs" or "pred"[2] (Chapter 10).
COMMAND	Character(len=max_string_len[3])	Operating-system command used to initiate a process-model run.
COMMANDID	Character(len=12)	Identifier for COMMAND.
COVMATRIX	Character(len=12)	Identifier for a variance-covariance matrix for a group of observations, predictions, or prior-information equations.
DERIVATIVES_INTERFACE	Character(len=max_string_len[3])	Name (or pathname) of a derivatives interface file (Chapters 7 and 13).
EQUATION	Character(len=max_string_len[3])	Right-hand side of an equation (Chapter 9) defining how one value is to be calculated from one or more other values. For observations and predictions, "_" can be used to indicate the value is extracted directly from a model-output file. The default is "_".
GROUPNAME	Character(len=12)	Identifier for a group of observations, predictions, prior-information equations, or parameters.
INSTRUCTIONFILE	Character(len=max_string_len[3])	Filename (or path ending in filename) of an instruction file (Chapter 8).
MATRIXFILE	Character(len=max_string_len[3])	Filename (or path ending in filename) of a file containing one or more matrices.
MEASSTATFLAG	Character(len=6)	Flag indicating what MEASSTATISTIC is (Table 10-2); valid values are "CV", "SD", and "VAR"[2].
MEASSTATISTIC	Character(len=max_string_len[3])	Statistic from which a variance is to be calculated (Table 10-2).

89

Keyword or Column name	Data type	Definition and, if applicable, valid values
MODINFILE	Character(len=max_string_len[3])	Filename (or path ending in filename) of a model-input file (Chapter 8).
MODOUTFILE	Character(len=max_string_len[3])	Filename (or path ending in filename) of a model-output file (Chapter 8).
NMATRICES	Integer	Number of matrices to be read from a file.
NONDETECT	Logical[1]	True indicates the value entered for OBSVALUE is a detection limit; false indicates the value is a measured value.
OBSNAME	Character(len=lendnam[3])	Identifier for an observation.
OBSVALUE	Floating-point	Observed value.
OPERATINGSYSTEM	Character(len=20)	Name of operating system on which application runs; valid values are: "windows" "linux", "dos", and "unix"[2]. The default is "windows".
PARALLEL	Logical[1]	If true, parallel processing is enabled (Chapter 12).
PARAMNAME	Character(len=12)	Parameter name.
PLOTSYMBOL	Integer	Integer to be associated with a group of observation, predictions, or prior-information equations.
PREDNAME	Character(len=lendnam[3])	Identifier for a prediction.
PRIORINFOVALUE	Floating-point	Prior-information value for a prior-information equation (Chapter 11).
PRIORNAME	Character(len=lendnam[3])	Identifier for a prior-information equation (Chapter 11).
PURPOSE	Character(len=12)	Purpose for COMMAND; valid values are "forward", "forward&der", and "derivatives"[2].
REFVALUE	Floating-point	Reference value for a prediction.
RUNNERDIR	Character(len=max_string_len[3])	File-system name of directory in which runner program JRUNNER runs (Chapter 12).
RUNNERNAME	Character(len=20)	Identifier for an instance of RUNNERDIR (Chapter 12).
RUNTIME	Floating-point	Expected execution time to run process model in a runner directory (Chapter 12).
STARTVALUE	Floating-point	Parameter value at start of analysis by an application.
STATFLAG	Character(len=6)	Flag indicating what STATISTIC is (Table 10-1).

Keyword or Column name	Data type	Definition and, if applicable, valid values
STATISTIC	Floating-point	Statistic from which a weight is to be calculated (Table 10-1); valid values are "cv", "sd", "var", "wt", and "sqrwt"[2].
TEMPLATEFILE	Character(len=max_string_len[3])	Filename (or path ending in filename) of a template file (Chapter 8).
TIMEOUTFACTOR	Floating-point	Multiplier for RUNTIME, where the product is used to determine if a process-model run is overdue (Chapter 12).
TRANSFORM	Logical[1]	If true, parameter is to be log-transformed for analysis by the application.
USEFLAG	Logical[1]	If true, the observation, prediction, or prior-information equation is to be used in the analysis.
VERBOSE	Integer	Flag used to determine verbosity of application output. Valid values are in the range 0 to 5.
VERBOSERUNNER	Integer	Flag used to determine verbosity of output of the jrunner program. Valid values are in the range 0 to 5.
WAIT	Floating-point	Wait interval, in seconds, used by subroutine PLL_WAIT (Chapter 12) before returning.
WTMULTIPLIER	Floating-point	Multiplier used to multiply weights of all members of a group of observations, prediction, or prior-information equations.
WTOSCONSTANT	Floating-point	A constant (η in the equation below) that can be added to control how much small values of $\tilde{y}_i$ influence the adjustable weight calculated for observation i (ω_i), where cv_i is the coefficient of variation, $\tilde{y}_i$ is the observed or simulated value depending on the current iteration, and $$\omega_i = \frac{1}{\left(cv_i \tilde{y}_i + \eta\right)^2}.$$

[1]Any logical variable is assigned as true if the value in the input block is "true", "yes", "t", or "y"; it is false if the value is "false", "no", "f", or "n" (recognition is case-insensitive).

[2]Recognition of text is case-insensitive.

[3]Variable is defined in the Global Data Module.

Naming Convention

User-specified names for parameters, observations, and predictions need to conform to a naming convention developed for the JUPITER API. The naming convention is designed to enable the Equation Module to distinguish easily between numbers and names when evaluating equations. Applications that use the JUPITER API can invoke subroutine UTL_CHECK_NAMES (Appendix D) to ensure compliance with the convention. The following two rules comprise the naming convention:

Rule 1. The first character needs to be a letter of the English alphabet.
Rule 2. All characters after the first letter need to be a letter, digit, or member of the set: "_", ".", ":", "&", "#", "@" (underscore, dot, colon, ampersand, number sign, at symbol).

Names of parameters are limited to 12 characters. Names of observations and predictions are limited to LENDNAM characters, where LENDNAM is defined in the Global Data Module as 20.

Chapter 17: Data-Exchange Files

By Eileen P. Poeter, Edward R. Banta, and Mary C. Hill

To facilitate rapid development of new methods using the JUPITER API and innovative application of those methods, standard formats are needed for files communicating data among applications. This need is addressed using data-exchange files.

Data-exchange files may be either explicitly or implicitly supported by the JUPITER API.

Explicit support of a data-exchange file means that the JUPITER API includes a subroutine for writing the data-exchange file, and possibly a subroutine for reading the data-exchange file.

Three functions of the Utilities Module (UTL_DX_OPEN, UTL_DX_CLOSE, and UTL_DX_GETUNIT) can be used to store or retrieve unit numbers associated with data_exchange files and open and close the files. These functions are designed to accept an extension associated with any data-exchange file in their argment lists.

When UTL_DX_OPEN is invoked with an extension that is not explicitly supported, then support for that file is implicit. Writing and (or) reading of implicitly supported data-exchange files is the responsibility of the application code.

Data-Exchange File Names, Content, and Format

JUPITER data-exchange files facilitate exchange of data among JUPITER application codes and post-processing software. These are ASCII files that contain data with little or no explanatory text because they are intended to be read by another computer program. Most data-exchange files are structured in a column format; for these files, the first line of the file contains column headings enclosed in double quotation marks. Some of the data-exchange file formats do not conform to a simple column format and are described separately. A few files have more than one heading-line.

The names of data-exchange files are formed from an application- or user-defined file name base and an extension associated with a particular data-exchange file type. The extension may be one of the extensions defined by the API or it may be application-defined. The first character of the extension conventionally is an underscore character but is not required to be. Many of the API-defined data-exchange files with an extension comprise of an underscore and two letters describe graphs, and the letters are chosen to coordinate with the y-axis variable [first letter] and the x-axis variable [second letter]). Typically the columns in the files are ordered with the x-axis column first because plotting software typically expects the values for the x-axis first. For example, the _os file is intended to create a graph of observed values against simulated equivalents; the first column of the file is the simulated equivalent and the second column if the observed value.

Data-exchange files are opened and assigned a unit number by invoking function UTL_DX_OPEN, and they are closed by invoking UTL_DX_CLOSE. The UTL_DX_GETUNIT function can be used to determine if a particular type of data-exchange file is open, and, if it is open, the unit number associated with the file.

An overview of all data-exchange files is provided in Table 17-1. The contents of files pertaining to residual analysis are presented in Table 17-2 and Table 17-3; files pertaining to sensitivity analysis are presented in Table 17-4; those related to regression analysis are tabulated in Table 17-5; files associated with prediction analysis are tabulated in Table 17-6; and a file including items for overall model evaluation is tabulated in Table 17-7.

Chapter 17: Data-Exchange Files

Table 17-1. Content of explicitly supported data-exchange files, in alphabetical order by file extension.
[--, no entry; ≥, greater than or equal to]

File extension	Content	Type	Written by:	Read by:
_dm	Information related to model structure, fit, and parsimony	Model evaluation	STA_UEV_DX_WRITE_DM	STA_UEV_DX_READ_DM
_gm	Dependent and (or) prior information groups and members	Residual analysis	DEP_DX_WRITE_GM (expects unit IUGM to be open for writing)	--
_mc	Parameter correlation matrix	Parameter analysis	UTL_DX_WRITE_MCMV	UTL_DX_READ_MCMV
_mv	Parameter variance/covariance matrix	Parameter analysis	UTL_DX_WRITE_MCMV	UTL_DX_READ_MCMV
_nm	Weighted residuals against probability plotting positions, which are also called standard normal statistics	Residual analysis	STA_UEV_DX_WRITE_NM	--
_os	Observations (dependents and prior information) against unweighted simulated equivalents	Residual analysis	DEP_UEV_DX_WRITE_OS (expects unit IUOS to be open for writing)	DEP_UEV_DX_READ_OS
_p	Predictions	Predictive analysis	DEP_UEV_DX_WRITE_P	DEP_UEV_DX_READ_P and DEP_UEV_DX_READ_PNUM (to count predictions)
_pa	Parameter estimates listed by parameter, then by iteration	Parameter analysis	UTL_DX_WRITE_PA	--
_pasub	Parameter estimates listed by iteration, then by parameter	Parameter analysis	UTL_DX_WRITE_PA	--
_pcc	Large parameter correlation coefficients (≥0.85) formatted for presentation	Parameter analysis	UTL_DX_WRITE_PCC	--
_pr	Prior information equations	Residual analysis	PRI_UEV_DX_WRITE_PR	PRI_UEV_DX_READ_PR

Chapter 17: Data-Exchange Files

File extension	Content	Type	Written by:	Read by:
_r	Unweighted residuals (dependents and prior information)	Residual analysis	DEP_UEV_DX_WRITE_R (expects unit IUR to be open for writing)	--
_s1	One-percent scaled sensitivities	Sensitivity analysis	SEN_UEV_DX_WRITE_MATRIX (expects unit IUSM to be open for writing)	SEN_UEV_DX_READ_MATRIX (expects unit IUSM to be open for reading)
_sc	Composite scaled sensitivity	Sensitivity analysis	SEN_UEV_DX_WRITE_SEN	--
_sd	Dimensionless scaled sensitivities	Sensitivity analysis	SEN_UEV_DX_WRITE_MATRIX (expects unit IUSM to be open for writing)	SEN_UEV_DX_READ_MATRIX (expects unit IUSM to be open for reading)
_so	Sensitivity summary by observation	Sensitivity analysis	SEN_UEV_DX_WRITE_SEN	--
_ss	Sum of squared weighted residuals by iteration	Sensitivity analysis	DEP_UEV_DX_WRITE_SS	--
_su	Unscaled sensitivities	Sensitivity analysis	SEN_UEV_DX_WRITE_SEN or SEN_UEV_DX_WRITE_MATRIX (expects unit IUSM to be open for writing)	SEN_UEV_DX_READ_SU or SEN_UEV_DX_READ_MATRIX (expects unit IUSM to be open for reading)
_w	Weighted residuals (dependents and prior information)	Residual analysis	DEP_UEV_DX_WRITE_W	--
_ws	Simulated equivalents against weighted residuals (dependents and prior information)	Residual analysis	DEP_UEV_DX_WRITE_WS	DEP_UEV_DX_READ_WS
_wt	Weights (dependents)	Residual analysis	UTL_DX_WRITE_WT	UTL_DX_READ_WT
_wtpri	Weights (prior information)	Residual analysis	UTL_DX_WRITE_WT	UTL_DX_READ_WT
_ww	Weighted simulated equivalents against weighted observations (dependents and prior information)	Residual analysis	DEP_UEV_DX_WRITE_WW	--

Table 17-2. Residual analysis data-exchange file formats.
[--, no entry]

File extension	Graph or information	"Column Tags" – Surrounded by double quotes in header			
_gm	Any related to observations or residuals	GROUP NAME	MEMBER NAME	PLOT SYMBOL	
_nm	Weighted residual probability distribution	WEIGHTED RESIDUAL	STANDARD NORMAL STATISTIC	PLOT SYMBOL	OBSERVATION or PRIOR NAME
_os	Observed against simulated values	SIMULATED EQUIVALENT	OBSERVED or PRIOR VALUE	PLOT SYMBOL	OBSERVATION or PRIOR NAME
_pr	Prior-information	PRIOR NAME	PLOT SYMBOL	EQUATION[1] Note: equation will be in double quotes	NATIVE SPACE PRIOR VALUE
_r	Residuals	RESIDUAL	PLOT SYMBOL	OBSERVATION or PRIOR NAME	--
_w	Weighted Residuals	WEIGHTED RESIDUAL	PLOT SYMBOL	OBSERVATION or PRIOR NAME	--
_ws	Weighted residual against simulated equivalent	SIMULATED EQUIVALENT	WEIGHTED RESIDUAL	PLOT SYMBOL	OBSERVATION or PRIOR NAME
_ww	Weighted observed against weighted simulated	WEIGHTED SIMULATED EQUIVALENT	WEIGHTED OBSERVED or PRIOR VALUE	PLOT SYMBOL	OBSERVATION or PRIOR NAME

[1]See Chapter 9 for description of the equation format.

Table 17-3. Format for _wt and _wtpri data-exchange files for weight matrices.
[The _wt file type is for dependents, and the _wtpri file type is for prior information]

_wt or _wtpri file lines	Information
One line	The number of matrices in the file. If 1, only the compressed weight matrix is included. If 2, then the compressed weight matrix is followed by its square root.
One line	Label: COMPRESSEDMATRIX
One line	Three numbers; the last two are the same: Number of non-zero values in the matrix, number of dependents (or prior-information items), number of dependents (or prior-information items).
One line for each non-zero element of the weight matrix	Two numbers on each line, an integer followed by a real number: Matrix position (using column-major ordering), non-zero element of the weight matrix.
One line	Label: COMPRESSEDMATRIX
One line	Three numbers; the last two are the same: Number of non-zero values in the matrix, number of dependents (or prior-information items), number of dependents (or prior-information items).
One line for each non-zero element of the square root of the weight matrix	Two numbers on each line, an integer followed by a real number: Matrix position (using column-major ordering), non-zero element of the square root of the weight matrix.

Table 17-4. Sensitivity analysis data-exchange file formats.
[--, no entry]

File	Graph	"Column Tags" — Surrounded by double quotes in header					
_sc	Bar chart of CSS. Use the parameter names as labels.	PARAMETER NAME	COMPOSITE SCALED SENSITIVITY	LARGEST DSS FOR THE PARAMETER	OBSERVATION WITH LARGEST DSS	SECOND LARGEST DSS FOR THE PARAMETER	OBSERVATION WITH SECOND LARGEST DSS
_so	Bar chart of leverage statistics for each observation.	OBSERVATION NAME	PLOT SYMBOL	LEVERAGE	LARGEST DSS	PARAMETER WITH LARGEST DSS	--
_sd	Bar chart of DSS. Use the observation numbers as labels. One graph for each parameter	OBSERVATION NAME	PLOT SYMBOL	ParamName Column of DSS for this parameter	ParamName Column of DSS for this parameter	Total number of columns: number of parameters +2	--
_s1	None. Limited to comparisons of dependent variables of the same type.	OBSERVATION NAME	PLOT SYMBOL	ParamName Column of 1-percent scaled-sensitivities for this parameter	ParamName Column of 1-percent scaled-sensitivities for this parameter	Total number of columns: number of parameters +2	--
_su	None. This file is used by other programs for calculations.	OBSERVATION NAME	PLOT SYMBOL	ParamName Column of unscaled sensitivities for this parameter	ParamName Column of unscaled sensitivities for this parameter	Total number of columns: number of parameters +2	--

Table 17-5. Parameter analysis data-exchange file formats.

File	Graph	"Column Tags" – Surrounded by double quotes in header
_mc	Parameter Correlation Matrix	ParamName ParamName ParamName ... Columns of the parameter correlation matrix. The order of parameters for rows is the same as the columns. — Total number of columns = NPE[1] = The total number of rows
_mv	Parameter Variance-Covariance Matrix	ParamName ParamName ParamName ... Columns of the parameter variance-covariance matrix with the order of parameters for rows the same as labeled for columns — Total number of columns = NPE[1] = The total number of rows
_pa	Parameter estimates by iteration	ITERATION — ESTIMATE
_pasub	None; designed for convenient preparation of input	ITERATION — ESTIMATE
_pcc	Large parameter correlation coefficients (≥0.85) formatted for presentation	PARAMETER — PARAMETER — CORRELATION
_ss	Sum-of-squared residuals	ITERATION — SSWR-(OBSERVATIONS ONLY) — SSWR-(PRIOR INFORMATION ONLY) — SSWR-(TOTAL) — # OBSERVATIONS INCLUDED

[1]NPE is number of adjustable parameters.

Table 17-6. Format for _p data-exchange file for predictions.

File	Graph / Information	"Column Tags" – Surrounded by double quotes in header		
_p	Predictions	PREDICTED VALUE	PLOT SYMBOL	PREDICTION NAME

Table 17-7. Format for the _dm data-exchange file for model evaluation.

Label	Example data
"MODEL NAME"	"ex1fullprior"
"MODEL LENGTH UNITS"	"M"
"MODEL MASS UNITS"	"NA"
"MODEL TIME UNITS"	"D"
"NUMBER ESTIMATED PARAMETERS"	5
"ORIGINAL NUMBER EXTIMATED PARAMETERS"	6
"TOTAL NUMBER PARAMETERS"	9
"NUMBER OBSERVATIONS INCLUDED"	35
"NUMBER OF OBSERVATIONS PROVIDED"	35
"NUMBER PRIOR"	3
"REGRESSION CONVERGED"	"YES"
"CALCULATED ERROR VARIANCE:"	1.48097002396
"STANDARD ERROR OF THE REGRESSION:"	1.21695111815
"MAXIMUM LIKELIHOOD OBJECTIVE FUNCTION – DEPENDENTS (MLOFD):"	-29.31161
"MAXIMUM LIKELIHOOD OBJECTIVE FUNCTION – DEPENDENTS AND PRIOR (MLOFDP):"	-27.27987
"AICc (MLOFD + AICc PENALTY): "	-17.24264
"BIC (MLOFD + BIC PENALTY): "	-11.53487
"HQ (MLOFD + HQ PENALTY): "	-16.62708
"KASHYAP (MLOFD + KASHYAP PENALTY): "	11.80662
"LN DETERMINANT OF FISHER INFORMATION MATRIX"	32.53087
"RN2 DEPENDENTS:"	0.9890975
"RN2 DEPENDENTS AND PRIOR:"	0.9920318
"NUMBER OF ITERATIONS"	4

Chapter 18: Code and Example Applications
By Edward R. Banta

The Fortran 90 source code for the JUPITER API and example applications with input and output files can be downloaded from the URL:
<http://water.usgs.gov/nrp/gwsoftware/jupiter/jupiter_api.html>. When extracted, the downloaded file generates several directories in the following directory structure:

```
jupiter_api
    bin
    doc
    mod
    obj
    src
        api_modules
        group_example
        jrunner
        sensitivity_example
    test
        group_example
        sensitivity_example
            runner1
            runner2
```

In the distribution file for the Windows operating system, the "bin" directory contains executable files for the programs GROUP_EXAMPLE, JRUNNER, SENSITIVITY_EXAMPLE, and MODFLOW-2000.

The "doc" directory contains supplemental documentation files, if any are needed. The "mod" directory is empty but is provided for storage of compiled module files.

The "obj" directory is empty but is provided for storage of object files.

The "src" directory contains source doc files in four subdirectories. The "api_modules" subdirectory contains source-code files for the modules that comprise the JUPITER API. The "group_example" subdirectory contains a source-code file for the GROUP_EXAMPLE application. The "jrunner" subdirectory contains source-code files for the JRUNNER application. The "sensitivity_example" subdirectory contains source-code files for the SENSITIVITY_EXAMPLE application.

The "test" directory has two subdirectories, which contain test-case input and output files for the GROUP_EXAMPLE and SENSITIVITY_EXAMPLE applications. The "runner1" and "runner2" subdirectories are runner directories populated with input files that allow SENSITIVITY_EXAMPLE to be run in parallel mode when JRUNNER is run in those directories.

Source code for the JUPITER API

Source-code files defining Fortran modules that are part of the JUPITER API are in the "api_modules" subdirectory. These modules provide all capabilities defined as part of the

101

JUPITER API, but they do not constitute programs themselves. Each source-code file contains one API module, as shown in the following table:

File name	JUPITER API Module name	USE statement syntax
bas f90	Basic Module	USE BASIC
dep.f90	Dependents Module	USE DEPENDENTS
eqn.f90	Equation Module	USE EQUATION
gdt.f90	Global Data Module	USE GLOBAL_DATA
mio.f90	Model Input-Output Module	USE MODEL_IO
pll.f90	Parallel-Processing Module	USE PARALLEL_PROCESSING
pri f90	Prior-Information Module	USE PRIOR_INFORMATION
sen f90	Sensitivity Module	USE SENSITIVITY
sta f90	Statistics Module	USE STATISTICS
typ.f90	Datatypes Module	USE DATATYPES
utl.f90	Utilities Module	USE UTILITIES

Compliance with Fortran-90 standard

The source code for the JUPITER API complies with the ANSI Fortran-90 standard (American National Standards Institute, 1992) with two exceptions: (1) Subroutine UTL_GETARG uses an extension to the standard (intrinsic subroutine GETCL) to obtain command-line arguments, and (2) subroutine UTL_SYSTEM uses an extension to the standard (intrinsic subroutine SYSTEM) to initiate an operating-system command.

Some compilers may support an alternative to GETCL or an alternative syntax to retrieve command-line arguments; in this case, UTL_GETARG needs to be modified. If a compiler supports no such capability, applications using the JUPITER API need to avoid calling UTL_GETARG.

Similarly, some compilers may have an alternative subroutine or syntax for initiating an operating-system command. In this case, UTL_SYSTEM needs to be modified to conform to the required syntax. For an application to make full use of the JUPITER API, a compiler that supports the ability to initiate an operating-system command is needed.

Example Application: GROUP_EXAMPLE

This simple example illustrates the use of three subroutines of the Utilities Module (Appendix D) for reading and storing data in input blocks. It uses the LLIST data type, which is one of the input-block data types described in Chapter 4. UTL_READBLOCK is called twice to read two input blocks: first to define groups (Chapter 3) and second to define members of the groups. UTL_GROUPLIST is then called to use the attributes associated with groups in the first input block as defaults for the group members. Finally, UTL_FILTERLIST is called several times to populate arrays in which each element contains an attribute value for one member. Program GROUP_EXAMPLE follows. The use of the NOCOL array, which has a dimension of 0, in the calls to UTL_READBLOCK causes the input blocks to have no default column order.

```
PROGRAM GROUP_EXAMPLE
USE DATATYPES
USE GLOBAL_DATA, ONLY: LENDNAM
USE UTILITIES
IMPLICIT NONE
!
INTEGER :: I,IERR,INUNIT,IOUT,KGPS,KOBS,MORE
TYPE (LLIST), POINTER :: LLPTRG, LLPTRM, LTAIL
CHARACTER (LEN=5),         DIMENSION(5) :: SFLAG
DOUBLE PRECISION,         DIMENSION(5) :: STATI, OBSVAL
CHARACTER (LEN=LENDNAM), DIMENSION(5) :: OBSNAM
CHARACTER (LEN=9),         DIMENSION(5) :: GROUPNAM
CHARACTER(LEN=40),        DIMENSION(0) :: NOCOL
!
!   Format statements
100 FORMAT('OBSNAME',2X,'GROUPNAME',2X,'  OBSVALUE  ',  &
           2X,'STATFLAG',2X,'STATISTIC', &
           /,'-------',2x,'---------',2X,'------------',  &
           2X,'--------',2X,'---------')
120 FORMAT(A7,2X,A9,2X,G12.5,4X,A,3X,G9.2)
!
NULLIFY(LLPTRG,LLPTRM,LTAIL)
INUNIT=7
IOUT=8
KGPS=0
KOBS=0
!
OPEN(INUNIT,FILE='gp.in',status='OLD')
OPEN(IOUT,FILE='gp.out',status='REPLACE')
!
!   Read input blocks
CALL UTL_READBLOCK(0,'OBSERVATION_GROUPS',NOCOL,INUNIT,IOUT,   &
                   'GROUPNAME',.TRUE.,LLPTRG,LTAIL,KGPS)
CALL UTL_READBLOCK(0,'OBSERVATION_DATA',NOCOL,INUNIT,IOUT,   &
                   'OBSNAME',.TRUE.,LLPTRM,LTAIL,KOBS)
!
!   Apply group attributes as defaults for group members
CALL UTL_GROUPLIST('DEFOBS        ',LLPTRG,IOUT,LLPTRM,KGPS,KOBS)
!
!   Populate arrays
CALL UTL_FILTERLIST(LLPTRM,IOUT,'OBSNAME',5,IERR,OBSNAM,MORE)
CALL UTL_FILTERLIST(LLPTRM,IOUT,'OBSVALUE',5,IERR,OBSVAL,MORE)
CALL UTL_FILTERLIST(LLPTRM,IOUT,'STATFLAG',5,IERR,SFLAG,MORE)
CALL UTL_FILTERLIST(LLPTRM,IOUT,'STATISTIC',5,IERR,STATI,MORE)
CALL UTL_FILTERLIST(LLPTRM,IOUT,'GROUPNAME',5,IERR,GROUPNAM,MORE)
!
!   Write output
WRITE(IOUT,100)
DO I=1,KOBS
  WRITE(IOUT,120)OBSNAM(I),GROUPNAM(I),OBSVAL(I),SFLAG(I),STATI(I)
ENDDO
STOP
END PROGRAM GROUP_EXAMPLE
```

Program GROUP_EXAMPLE reads input from file gp.in:

```
BEGIN Observation_Groups KEYWORDS
  GROUPNAME=HeadObs
    STATFLAG=var
    STATISTIC=1.5
  GROUPNAME=FlowObs
    STATFLAG=cv
    STATISTIC=0.08
END Observation_Groups

BEGIN Observation_Data KEYWORDS
  OBSNAME=Head1  GROUPNAME=HeadObs   OBSVALUE=101.8
  OBSNAME=Head2  GROUPNAME=HeadObs   OBSVALUE=98.6     STATISTIC=1.0
  OBSNAME=Head3  GROUPNAME=HeadObs   OBSVALUE=149.3
  OBSNAME=Flow1  GROUPNAME=FlowObs   OBSVALUE=120.0
  OBSNAME=Flow2  GROUPNAME=FlowObs   OBSVALUE=65.0
END Observation_Data
```

and writes the following output to file gp.out:

```
OBSNAME   GROUPNAME    OBSVALUE      STATFLAG   STATISTIC
-------   ---------    ------------  --------   ---------

Head1     HeadObs      101.80        var        1.5
Head2     HeadObs      98.600        var        1.0
Head3     HeadObs      149.30        var        1.5
Flow1     FlowObs      120.00        cv         0.80E-01
Flow2     FlowObs      65.000        cv         0.80E-01
```

Example Application: SENSITIVITY_EXAMPLE

The source-code files for this example JUPITER application are in the "sensitivity_example" subdirectory under the "src" directory of the JUPITER API distribution. The file "sensitivity_example.f90" contains a main program unit and the file "sensitivity_example_mod.f90" contains an associated application module. These files, in combination with source-code files in the "api_modules" subdirectory, can be compiled and linked to generate the SENSITIVITY_EXAMPLE application. When compiling the SENSITIVITY_EXAMPLE application, all modules of the JUPITER API must be compiled and made available before "sensitivity_example_mod.f90" and "sensitivity_example.f90" are compiled.

SENSITIVITY_EXAMPLE is designed to produce simulated equivalents for observations and sensitivities of those simulated equivalents to parameters. It can calculate perturbation sensitivities by either a forward-difference or a central-difference method, and it can obtain sensitivities directly from a model designed to calculate sensitivities internally. SENSITIVITY_EXAMPLE can use any combination of these methods in a single run. It also can use simulated equivalents generated during a model run that also generates model-calculated sensitivities. SENSITIVITY_EXAMPLE expects an input file as an argument on the command line. The "sensitivity_example_input.txt" file can be used as this input file. The main output file is "SENSITIVITY_EXAMPLE.out". When run using "sensitivity_example_input.txt" as input, a data-exchange file of type _su (Chapter 17) named "tc1_sen_example_out._su" also is written; this file contains unscaled sensitivities of simulated equivalents to observations with respect to all adjustable parameters.

104

SENSITIVITY_EXAMPLE reads the input blocks shown in Table 18-1 and other data from "sensitivity_example_input.txt" and other input files.

Table 18-1. Input blocks read by SENSITIVITY_EXAMPLE.
[See Chapter 16 for additional description of all input blocks except Control_Data.]

Blocklabel for input block	Required or optional	Content	Chapter[1]
Options	Optional	Options recognized by JUPITER API.	7
Control_Data	Required	Settings and options recognized by SENSITIVITY_EXAMPLE.	--
Model_Command_Lines	Required	Command(s) for initiating model runs.	7
Parameter_Groups	Optional	Attributes applicable to groups of parameters.	--
Parameter_Data	Required	Attributes applicable to individual parameters.	
Parameter_Values	Optional	Parameter names and values.	
Observation_Groups	Optional	Attributes applicable to groups of observations.	10
Observation_Data	Required	Attributes applicable to individual observations.	10
Matrix_Files	Optional	Variance-covariance matrices for groups of observations.	7, 10
Model_Input_Files	Required	Names of model-input and template files.	8
Model_Output_Files	Required	Names of model-output and instruction files.	8
Parallel_Control	Optional	Settings and options to enable parallel processing.	12
Parallel_Runners	Optional	Descriptions of runners used for parallel processing.	12

[1] No chapter is identified if the blocklabel is not supported by programming provided as part of the JUPITER API.

Example Application: JRUNNER

When using the SENSITIVITY_EXAMPLE program (documented above) in parallel mode, a runner program must be running in each runner directory. This section documents JRUNNER, which can be used as the runner program. Note that JRUNNER is not specific to SENSITIVITY_EXAMPLE but is general in that it can be used with any application that is designed to have a parallel-processing mode and uses the JUPITER API to perform the "Adapt Parameter Valu es", "Execute Process Model", and "Extract Values from Model Output" tasks. Design of a parallel-processing capability in an application is beyond the scope of this report. However, developers using the JUPITER API for applications that require many model runs (which could be made concurrently) are encouraged to familiarize themselves with the main program unit of SENSITIVITY_EXAMPLE to see one method for approaching the parallelization problem. See Chapter 12 for descriptions of the parallel-processing methodology and JRUNNER. When run in the parallel-processing mode, SENSITIVITY_EXAMPLE is considered the dispatcher program.

The code for JRUNNER is contained in two application-specific source-code files: jrunner.f90 and jrunner_mod.f90. Jrunner.f90 contains the main program unit for JRUNNER, and jrunner_mod.f90 contains the application module. The program units in these files use the Global Data, Datatypes, Utilities, Parallel Processing, and Model Input-Output Modules; these modules must be compiled and available before JRUNNER can be compiled. The source-code files are in the JUPITER API distribution file.

To run the test case in parallel mode (1) edit the sensitivity_example_input.txt file in the test directory to change "PARALLEL=no" to "PARALLEL=yes" in the PARALLEL_CONTROL input block, (2) start JRUNNER in the runner1 and runner2 subdirectories, then (3) start SENSITIVITY_EXAMPLE using sensitivity_example_input.txt as the input file.

When VERBOSERUNNER in the PARALLEL_CONTROL input block for a dispatcher program, in this case SENSITIVITY_EXAMPLE, is assigned a value larger than 3, JRUNNER writes a file that lists dependent names and corresponding extracted values for each run. The file is opened as an implicitly supported data-exchange file (Chapter 17) and is named "run#####._ext", where "#####" is replaced by the run number, padded with zeroes as needed.

References Cited

Ahlfeld, D.P., and Mulligan, A.E., 2000, Optimal management of flow in groundwater systems: San Diego, California, Academic Press, 185p.

American National Standards Institute, Inc., 1992, American National Standard for Programming Language – Fortran – Extended, ANSI X3.198-1992, 369 p.

Babendreier, J. E., 2004, Calibration, optimization, and sensitivity and uncertainty algorithms application programming interface (COSU-API): International Workshop on Uncertainty, Sensitivity, and Parameter Estimation for Multimedia Environmental Modeling, Rockville, MD, August 19 - 21, 2003. U.S. Nuclear Regulatory Commission, Washington, DC, A1-A54.

Bicknell, B.R., Imhoff, J.C., Kittle, J.L., Jr., Donigian, A.S., Jr., and Johanson, R.C., 1997. Hydrological Simulation Program--Fortran, User's manual for version 11. U.S. Environmental Protection Agency, National Exposure Research Laboratory, Athens, Ga., EPA/600/R-97/080, 755 pp.

Brockwell, P.J., and Davis, R.A., 1987, Time series—Theory and methods: New York, Springer-Verlag, 519 p.

Burnham, K.P., and Anderson, D.R., 2002, Model selection and multi-model inference: A practical information-theoretic approach: Springer-Verlag, New York, 488 p.

Cook, R.D., and Weisberg, S., 1982, Residuals and influence in regression: New York, Chapman and Hall.

Cook, R.D., and Weisberg, S., 1999, Applied regression including computing and graphics: New York, Wiley.

Dennis, J.E., and Schnabel, R.B., 1996, Numerical methods for unconstrained optimization and nonlinear equations: Philadelphia, Pennsylvania, USA, Society for Industrial and Applied Mathematics, Classics in Applied Mathematics 16, 378p.

Doherty, John, 2004, PEST—Model-Independent Parameter Estimation, User Manual, 5th ed.: Corinda, Australia, Watermark Numerical Computing, 336 p.

Draper, N.R., and Smith, H, 1998, Applied regression analysis (3rd ed.): New York, Wiley.

Harbaugh, A. W., Banta, E. R., Hill, M. C., and McDonald, M. G., 2000. MODFLOW-2000, The U.S. Geological Survey modular ground-water model – Users guide to modularization concepts and the ground-water flow process. U.S. Geological Survey Open-File Report 00-92, 121 pp.

Harbaugh, A.W, 2005, MODFLOW-2005, the U.S. Geological Survey modular ground-water model—the ground-water flow process: U.S. Geological Survey Techniques and Methods, book 6, chap. A16, variously paginated.

Hill, M.C., Banta, E.R., Harbaugh, A.W., and Anderman, E.R., 2000, Modflow-2000, the U.S. Geological Survey modular ground-water model—User guide to the Observation, Sensitivity, and Parameter-Estimation Processes and three post-processing programs: U.S. Geological Survey Open-File Report 00–184, 209 p.

Hill, M.C., and Tiedeman, C.R., in press, Effective groundwater model calibration, with analysis of sensitivities, predictions, and uncertainty: New York, Wiley and Sons.

Leavesley, G.H. and Stannard, L.G., 1995. The precipitation-runoff modeling system – PRMS, in Singh, V.P., ed., Computer models of watershed hydrology. Highlands Ranch, CO, Water Resources Publications, Ch. 9, 281-310.

Menke, W., 1984, Geophysical Data Analysis, Discrete Inverse Theory: Elsevier, New York.

Parker, R.L., 1994, Geophysical inverse theory: Princeton (NJ), Princeton University Press.

References Cited

Poeter, E.P., and Anderson, D., 2005, Multi-model ranking and inference in groundwater modeling: Ground Water v. 43 no. 4, p. 597-605.

Poeter, E.P., Hill, M.C., Banta, E.R., Mehl, Steffen and Christensen, Steen, 2005, UCODE_2005 and six other computer codes for universal sensitivity analysis, calibration, and uncertainty evaluation: U.S. Geological Survey Techniques and Methods, book 6 chap. A11, 283 p.

Saltelli, A., Chan, K., and Scott, E.M., 2000, Sensitivity Analysis: New York, Wiley.

Saltelli, A., Tarantola, S., Campolongo, F., and Ratto, M., 2004, Sensitivity Analysis in Practice. New York: Wiley.

Seber, G.A.F., and Wild, C.J., 1989, Nonlinear Regression. New York: Wiley.

Shapiro, S.S., and Francia, R.S., 1972, An approximate analysis of variance test for normality: Journal of the American Statistical Association, v. 67, p. 215–216.

Sun, N.-Z., 1994, Inverse problems in ground-water modeling. Boston (MA): Kluwer Academic Publishers.

Tarantola, A. 1994, Inverse problem theory. New York: Elsevier.

Tiedeman, C.R., Ely, D.M., Hill, M.C., and O'Brien, G.M., 2004, A method for evaluating the importance of system state observations to model predictions, with application to the Death Valley regional groundwater flow system. Water Resources Research 40, W12411. DOI: 10.1029/2004WR003313.

Tiedeman, C.R., Hill, M.C., D'Agnese, F.A., and Faunt, C.C., 2003, Methods for using groundwater model predictions to guide hydrogeologic data collection, with application to the Death Valley regional ground-water flow system: Water Resources Research 39(1). DOI: 10.1029/2001WR001255.

Zheng, Chunmiao, and P. Patrick Wang, 1999. MT3DMS, A modular three-dimensional multi-species transport model for simulation of advection, dispersion and chemical reactions of contaminants in groundwater systems; documentation and user's guide. U.S. Army Engineer Research and Development Center Contract Report SERDP-99-1, Vicksburg, MS, 202 pp. http://hydro.geo.ua.edu/mt3d/

Appendix A: Additional Input Instructions

By Edward R. Banta, John E. Doherty, Mary C. Hill, and Eileen P. Poeter

Except for the input blocks described in Chapter 3 and in the chapters for each module, this Appendix includes instructions for all input defined as part of the API. This includes instructions for:

- Template files used to create model-input files that reflect updated parameter values. The names of the Template files are listed in the Model_Input input block of the Model Input-Output Module of Chapter 8.
- Instruction files used to read values from model-output files. The names of the instructions files are listed in the Model_Output input block of the Model Input-Output Module of Chapter 8.
- Derivatives Interface file used to read model-calculated derivatives from model-generated files. Use of the derivatives is determined by the application. The name of the derivatives interface file is listed in the Options input block of the Basic Module (Chapter 7).
- Matrices.

Template Files

When a JUPITER application runs a model, it generally needs to "adapt" model-input files to reflect current parameter values as determined by the JUPITER application. This is accomplished using "template files." Of the many input files a model may use, template files are needed only for input files that contain parameters with values manipulated by the JUPITER application.

An Example

A template file is a replica of a model-input file except that spaces occupied by numbers affected by manipulated parameter values are replaced by a sequence of characters. The application substitutes numbers for the sequence of characters to create a new model-input file.

Consider the model-input file shown in Figure A-1; this file supplies data to a program that computes the "apparent resistivity" on the surface of a layered half-space for different surface electrode configurations. Suppose the modeler wants to use this program (that is, process model) to estimate the properties for each of three half-space layers from apparent resistivity data collected on the surface of the half-space. The parameters to be estimated are the resistivity and thickness of the upper two layers and the resistivity of the third (for mathematical analysis, its thickness is assumed as infinite). A suitable template file appears in Figure A-2.

```
MODEL-INPUT FILE
3, 15                                  no. of layers, no. of spacings
1.0, 1.0                                resistivity, thickness: layer 1
40.0, 20.0                             resistivity, thickness: layer 2
5.0                                    resistivity: layer 3
1.0                                    electrode spacings
1.47
2.15
3.16
4.64
6.81
10.0
14.9
21.5
31.6
46.4
68.1
100
149
215
```

Figure A-1. A model-input file.

```
jtf #
MODEL-INPUT FILE
3, 15                                  no. of layers, no. of spacings
#res1      #,#t1          #            resistivity, thickness: layer 1
#res2      #,#t2          #            resistivity, thickness: layer 2
#res3      #                           resistivity: layer 3
1.0                                    electrode spacings
1.47
2.15
3.16
4.64
6.81
10.0
14.9
21.5
31.6
46.4
68.1
100
149
215
```

Figure A-2. A template file corresponding to the model-input file of Figure A-1.

Appendix A: Additional Input Instructions

Parameter Spaces

As Figure A-2 shows, the first line of a template file contains the letters "jtf" followed by a space, followed by a single character; "jtf" stands for "JUPITER template file." The character following the space is the "parameter delimiter." In a template file, a "parameter space" is identified as the set of characters between and including a pair of parameter delimiters. Characters between these delimiters need to include the name of a parameter. When a JUPITER program writes a model-input file based on a template file, it replaces all characters between and including the parameter delimiters by a number representing the current value of the indicated parameter.

The parameter delimiter is selected by the user. It cannot be a character of the set [a-z], [A-Z] and [0-9], or any other character that appears elsewhere in the template file. Whenever a JUPITER program encounters the parameter delimiter in a template file, it assumes that it defines a parameter space. Good choices for the parameter delimiter are @, #, and !.

Setting the Parameter Space Width

Wider parameter spaces (up to a certain limit - see "How the Model Input-Output Module Fills a Parameter Space with a Number" section, below) often are advantageous because they allow numbers to be represented with greater precision. In many circumstances, such as the calculation of sensitivities using perturbation, precision is crucial. When sensitivities are calculated, enough precision is needed so that the number of significant digits of the difference between a parameter value and its perturbed value equals the number of significant digits desired for the sensitivity. It is good practice to define parameter spaces that take advantage of the model capabilities, so that parameter values can be provided to the model with the same precision with which they are calculated by JUPITER applications. If one parameter occurs more than once in model-input files, the parameter width used equals the smallest of the parameter widths defined.

Generally models read numbers from an input file in either of two ways: (1) from specified fields or (2) as a sequence of numbers each of which may be of any length. For specified fields, the model input instructions can be used to determine how a JUPITER API application needs to write numbers that are compatible with the model requirements. In Fortran, the latter method is often referred to as "free field input."

The Fortran code of Figure A-3 directs a program to read five real or double precision numbers. The first three are read using a format specifier; the last two are read as free format.

```
      READ(20,100) A,B,C
100   FORMAT(3F10.0)
      READ(20,*) D,E
```

Figure A-3. Formatted and free field input.

The relevant part of the input file may be as illustrated in Figure A-4.

```
     6.32  1.42E-05123.456789
34.567, 1.2E17
```

Figure A-4. Numbers read using the code of Figure A-3.

No whitespace or comma is needed between numbers that are read using a field specifier. Whitespace is used here to refer to a string of one or more space characters, or the beginning or end

111

of a line. The format statement labeled "100" in Figure A-3 specifies reading variable A from the first 10 positions on the line, reading variable B from the next 10 positions, and reading variable C from the 10 positions thereafter. When the model reads these numbers, characters outside the field being read are of no consequence. In contrast, when reading variables D and E, the whitespace or a comma is used to separate the numbers.

Suppose variables A to E are model parameters, and that a JUPITER application needs to transfer current values for these parameters to the model's input file. For convenience it is assumed that the parameters are named A, B, C, D, and E; the same names used in the program fragment of Figure A-3. A template fragment corresponding to the input file fragment of Figure A-4 is shown in Figure A-5. The parameter space for parameters A, B, and C is 10 characters wide. These three parameter spaces abut each other, as is allowed when using the format specifier of Figure A-3. If the parameter space for any of these parameters were greater than 10 characters in width in the template file, the resulting model-input file would not be readable by the model. Although not advisable, the parameter space width can be designated as less than the specified width if whitespace is used to produce the result such that only one number occupies each specified width.

```
#   A      ##     B    ##   C       #
#    D          #, #   E         #
```

Figure A-5. Fragment of a template file corresponding to Figure A-4.

For parameters D and E, two numbers need to appear in succession separated by whitespace or a comma; the length of the numbers is not defined by the model. In Figure A-5, the numbers are each allowed 13 spaces, which provides for considerable precision.

Template files commonly are constructed by starting with a model-input file. In model-input files, numbers read using free format are often written with trailing zeros omitted. In constructing the template file, expanding the parameter space for such numbers is advantageous. Ensure that numbers to be read in free format are bounded by whitespace, a comma, or the beginning or end of a line.

Numbers read using a specified field format need not occupy the full field width in a model-input file (for example, variable A in Figure A-4). For improved precision, it is better to expand the parameter space to take advantage of the entire field.

How the Model Input-Output Module Fills a Parameter Space with a Number

Subroutine MIO_ADA_WRITEFILES of the Model Input-Output Module is the subroutine responsible for the production of model-input files that are adapted to current parameter values.

MIO_ADA_WRITEFILES writes as many significant digits as the defined parameter space allows using arguments PRECPR and NOPNT. PRECPR may be 0 (meaning "single precision"), 1 (meaning "double precision using E format"), or 2 (meaning "double precision using D format"); it determines whether single- or double-precision protocol is observed in writing parameter values to model-input files. The precision of the number is affected only if the parameter width is greater than 13 characters. If PRECPR equals 0, exponents are represented by the letter "E." Also, if a parameter space is greater than 13 characters in width, only the last 13 spaces are used in writing the number representing the parameter value. Any remaining characters within the parameter space are left blank. If PRECPR equals 1 or 2, up to 23 characters can be used to represent a number. If PRECPR is 1, the letter "E" is used to represent exponents; if PRECPR is 2, the letter "D" is used to represent exponents. Double precision allows extremely large and extremely small numbers to be represented more precisely than single precision. Furthermore, double precision allows all numbers that can be represented in single precision to be represented more precisely. Double precision allows representation of numbers whose absolute values are too large to be represented in single precision. It also allows representation of numbers whose absolute values are too small to be represented as non-zero values in single precision. However, if the process model uses single-precision variables to read model input values, nothing is gained by writing the values in double precision.

If a model's input data fields are small, every effort is needed to obtain as much precision as possible from the limited parameter spaces available. Subroutine MIO_ADA_WRITEFILES may be able to gain one or more extra significant figures if it does not include a non-essential decimal point in a number. For example, if a parameter space is 5 characters wide and the current value of the parameter to which this field pertains is 10,234.345, the number will be written as "1.0e4" or as "10234" depending on whether the decimal point needs to be included. Similarly, if the parameter space is 6 characters wide, the number 106,857.34 can be represented as either "1.07e5" or "1069e2."

If NOPNT equals 1, subroutine MIO_ADA_WRITEFILES omits the decimal point if it can. Whether or not a number without a decimal point can be read by the model depends on the model. If the model is written in Fortran and numbers are read using free field input, or using a specified field ending in .0 such as "(F6.0)" or "(E8.0)", the decimal point is not needed. However specified fields such as "(F6.2)" and "(E8.2)" enforce power-of-10 scaling if a decimal point is absent from an input number. Hence if there is any doubt, specify NOPNT as 0 to ensure that all numbers written to model-input files include a decimal point, thus overriding point-location or scaling conventions implicit in some Fortran format specifiers.

If a parameter space is 13 characters wide or greater and PRECPR equals zero (implying single precision protocol), the decimal point is included regardless of the value of NOPNT; there is no gain obtained from leaving it out. Similarly, if PRECPR equals 1 or 2 (implying double precision protocol), MIO_ADA_WRITEFILES makes no attempt to omit a decimal point if the parameter space is 23 characters wide or more.

Table A-1 shows how the setting of NOPNT affects the representation of the number 12345.67. The table also shows how MIO_ADA_WRITEFILES writes a number to maximum precision.

Table A-1. Representations of the number 12345.67.

parameter space width (characters)	NOPNT = 0	NOPNT = 1
8	12345.67	12345.67
7	12345.7	12345.7
6	12346.	12346.
5	1.2e4	12346
4	1.e4	12e3
3	***	1e4
2	**	**

If a parameter is written to more than one field (which may be located in more than one template file) MIO_ADA_WRITEFILES writes the parameter value to all fields using the minimum parameter space width specified. In the wider spaces the number is right-justified. In this way a consistent parameter value is written to all spaces pertaining to the one parameter.

113

Appendix A: Additional Input Instructions

Preparing a Template File

Preparation of a template file generally is a simple procedure. Start with a copy of a model-input file. Insert the "jtf" header line at the beginning of the file and define a parameter space delimiter. Use a text editor to replace parameter values by their respective parameter space identifiers, whitespace, and parameter space delimiters.

Instruction Files

General

Of the possibly huge amount of information that a model may write to output file(s), a JUPITER application is normally interested in only a few numbers, such as numbers that correspond to field or laboratory measurements or predictions. These model-generated numbers are referred to as "dependents" in this document. For every model-output file containing dependents, an instruction file is needed to provide directions for reading the file. The model-output file and its instruction file are defined in the Model_Output_Files input block of the Model Input-Output Module.

An instruction file can be constructed in either of two ways: using the "instruction set" option or the "standard file" option. The "standard file" option is simpler, but only can be used to obtain numbers from files composed of initial header lines and subsequent columns of data. The "standard file" option is explained in the "Standard File" section. If the model-output file does not meet the requirements listed in the "Standard File" section, the "instruction set" option must be used. The "instruction set" option is explained in the "Instruction Set" section. Both types of instruction files require that a marker delimiter be specified and that dependent names be specified, so these are described first.

The Marker Delimiter

The first line of all instruction files needs to begin with the three letters "jif" (which stand for "JUPITER instruction file"), a single space, and a single character. The single character is the marker delimiter, which defines the beginning and end of a marker. A marker also could be called a search string; it is a sequence of characters that is the object of a search. A marker delimiter immediately precedes the first character of a marker and immediately follows the last character. For example, given "%" as the delimiter, %LAYER 1% identifies the marker "LAYER 1".

Appendix A: Additional Input Instructions

A marker delimiter cannot be any of the following characters because these characters have other roles in instruction files:

A B C D E F... Z
abcdef ... z
0 - 9
!
[
]
(
)
:
&

The character chosen for a marker delimiter also cannot occur within the text of any markers. Usually $, @, % or ~ are good choices for the marker delimiter.

Dependent Names

Dependents are linked to observations and predictions by names referred to as "ObsName" and "PredName" in Chapter 10. Dependent names need to be unique, with the number of characters not exceeding LENDNAM of the Global Data Module (Appendix C). In the JUPITER API distribution, LENDNAM equals 20. Name restrictions are as described in Chapter 10. Each dependent name can occur once in a set of the instruction files, except as noted below. This is because each dependent can be read only once from a model's output dataset. For standard files, these are the only aspects of dependent names that need to be considered.

The one name that can be repeated is the dummy dependent name "dum." When the "instruction set" option is used, "dum" can occur many times in an instruction file, and can be useful in navigating to a desired number. When "dum" is used, the Model Input-Output Module reads a number that is then ignored. The name "dum" cannot be used for any ObsName or PredName.

Standard File

If the model-output file from which dependent values are to be read meets the following requirements, the "standard file" option may be used:
1. A known, fixed number of lines preceeds the first line containing a dependent value to be read. This number of lines may be zero.
2. A known, fixed number of dependent values can be read from subsequent lines of the file, with exactly one dependent value per line.
3. Dependent values are in space-delimited columns, such that the dependent value to be read is in the same column on each line.
4. If numbers in more than one column are needed, the file can be listed twice in the model-output input block and a different column read each time.

Appendix A: Additional Input Instructions

Dependent values can be read from a model-output file that meets these requirements using an instruction file of the form:

```
jif marker-delimiter
StandardFile  Nskip  Readcolumn  Nread
Nread dependent names, one per line
```

where,

jif   is a keyword that defines the file delimiter that needs to be defined for all instruction files. The delimiter is not used in standard file instruction files, but needs to be defined.

StandardFile is a keyword. It is case insensitive.

Nskip   is the number of lines to skip at the beginning of the file and can equal 0 or any positive integer.

ReadColumn is the column of the file from which values are to be read. In this context, "column" refers to a whitespace-delimited text string, not to a position on a line determined by counting characters from the beginning of the line.

Nread   is the number of values, and therefore lines, to be read.

The entries on the line with keyword StandardFile are read in free format, so they need to be separated by whitespace or a comma. Instruction files are not allowed to have any comment lines or blank lines, even at the end of the file.

For example, five simulated values can be read from the _os data exchange files produced by MODFLOW-2000 as a standard file with zero lines to be skipped and the value to be read in the first column using a standard file containing:

```
jif @
StandardFile 0 1 5
Obs1
Obs2
Obs3
Obs4
Obs5
```

The numbers read will be compared with five observations defined in the Observation_Data input block corresponding to the names listed in this instruction file.

Execution time can be minimized by defining the names in the Observation_Data and Prediction_Data input blocks in the same order as they appear in the standardfile instruction file, which is the same as the input file being read. This is because a match is obtained by searching through the list read using the standardfile instructions for the items listed in the Observation_Data and Prediction_Data input files. If the order is the same, each search succeeds with minimum effort.

Appendix A: Additional Input Instructions

Example instruction file using the "standard file" option:

```
jif @
StandardFile 0 1 4
h1
h2
h3
h4
```

This instruction file will read the following model-output file and populate four dependents (named h1, h2, h3, and h4) with the values in column 1 (100.1, -0.7629E-04, -0.8516E-01, and 126.9) of the following model-output file:

```
  100.1         101.8       1   h1
 -0.7629E-04   -0.2899E-01  1   h2
 -0.8516E-01   -0.1289      1   h3
  126.9         128.1       1   h4
```

Instruction Set Needed for Non-Standard Files

If a model-output file from which dependent values are to be extracted does not conform to the rigid structure required when using the "standard file" option, the instruction file must contain detailed instructions of how to locate and read each dependent value. This section explains how to construct such an instruction file. The instructions available in the JUPITER API documented here are listed in Table A-2.

An extraction instruction can never be the first item on an instruction line. One of the following needs to come first: a primary marker, line advance item, or a continuation instruction. This is required for the MIO_EXT cursor to be located on a line on which one or more dependents may occur. If there is more than one dependent on a line of the model-output file, these dependents need to be read from left to right. Backward movement along a line or within a file is not allowed.

Extraction Type

Extractions can be categorized into three types, fixed, semi-fixed, and non-fixed, as shown in Table A-2. These types are discussed briefly here and in more detail in the "Example Instruction Files" section.

Fixed extraction instructions consist of two parts. The first part consists of the extraction name enclosed in square brackets; the second part consists of the range of columns (inclusive and separated by a semicolon) from which to extract the value. These parts cannot be separated by a space because a space always indicates the start of a new instruction unless the space lies within a marker. In this context, "column" refers to a position that can be occupied by one character on a line determined by counting characters (including whitespaces) from the beginning of the line.

Semi-fixed extraction instructions are similar to fixed extractions except that the extraction name is enclosed in parentheses (in contrast to square brackets for fixed instructions). However, the numbers do not locate the dependent exactly. A semi-fixed dependent instruction operates as follows. It sends the cursor to the first of the two columns indicated in the extraction instruction. If this column is occupied by a space, it searches to the right until it reaches either a non-space, which is taken to be the first character to be extracted, or the second listed column. If it reaches the second column without finding a non-space, an error condition arises. If it finds a non-space, it then searches to the right for the next blank or the end of the line, either of which is assumed to define the right side of the extraction. If the column identified by the first number is non-blank, the first character of the extraction is sought to the left and the last character of the extraction is sought to the right. The width of a number read through a semi-fixed abstraction can be longer than the range of columns cited in the semi-fixed extraction instruction.

Non-fixed extraction instructions do not include column numbers; the extraction name is enclosed by exclamation marks. The number is approached using other methods to navigate within a line of the process-model output file. A value to be extracted by non-fixed reading needs to be bounded by spaces, the beginning of a line, or the end of a line.

No blank lines are allowed within an instruction file.

Appendix A: Additional Input Instructions

Table A-2. Instructions available in the JUPITER API.
[All instruction lines need to start with something in the first column of the line. Each instruction needs to be bounded by the beginning of a line, a blank space, or the end of a line. No blank lines or comments are allowed in instruction files. In this table: # is any positive integer; name is "dum" or the name of an observation or a prediction; a, b are column numbers. Column numbers are obtained by counting from the left side of a line, starting at column 1.]

Instruction name	Format	Description	Example
Instructions that need to begin in the first column of a line in an instruction file			
Primary marker	Marker delimiter in the first column of a line in an instruction file, followed by a search string, followed by a marker delimiter.	Identifies a search string used to find a line within the process-model output file. The entire search string needs to be on one line.	%find me%
Line advance	l#, where l is a lower case L	Advances # lines down from the present line. When a file is opened the line is 0. Use l1 (line 1) to reach the first line.	l5
Continuation	& positioned in the first place in a line of an instruction file. Markers can not be broken across lines.	The instruction continues from the previous line.	&
Instructions that need to begin later in a line of an instruction file			
Secondary marker	Marker delimiter not in the first column of an instruction file, followed by a search string, and ending with a marker delimiter.	Identifies a search string to navigate within a line of the process-model input file. Secondary markers are preceded on the instruction file line by one of the instructions above.	l5 %find me%
Whitespace	w	Advance in a line to the last blank in the next set of blanks. Repeat as needed.	l5 w w
Tab	t#	Advance in a line to column #.	l5 t60
Extractions			
Fixed	[name]a:b	Read a number from a line starting at column a and ending at column b.	[dd1]1:10
Semi-fixed	(name)a:b	Read a number that may be only partially contained in the range extending from a to b, inclusive.	(dd1)2:8
Non-fixed	!name!	Read a number in free format. The number needs to be bounded by spaces, the beginning of a line, or the end of a line.	!dd1!

Appendix A: Additional Input Instructions

How a Model-Output File is Read

Dependents generally cannot be read from model-output files using the template concept employed for input file adaptation because most models do not produce an output file of identical structure on each model run. Instead, a list of instructions is used to find dependents in model-output files. Basically, a JUPITER application finds a dependent in a model-output file the same way a person does. A person looks down the file for something recognizable — a "marker." If this marker is properly selected, dependents usually can be subsequently found in a simple manner. For example, if the calculated stress at an elapsed time of 100 milliseconds is of interest, an instruction may be provided to read a model-output file looking for the marker:
```
STRESS CALCULATED AT FINITE ELEMENT NODES: ELAPSED TIME = 100 MSEC
```

The dependent may then be located, for example, by going down four lines and finding the number between character positions 23 and 30, or going down three lines and finding the fifth number on the line.

Markers can be of either primary or secondary type. The Model Input-Output Module scans a model-output file line by line looking for a primary marker. A secondary marker is sought on a single line, with the search proceeding from left to right. Subsequent reading or searching continues from the end of the secondary marker.

An Example Instruction File

Figure A-6 shows an output file written by the model whose input file appears in Figure A-1. Assume that the output file may have a variable number of lines preceding the line containing "electrode spacing." Suppose that parameters appearing in the template file of Figure A-2 are estimated by a JUPITER API application through minimization of the differences between apparent resistivities generated by the model and those provided by field measurements. Instructions that can be provided to the application to read each of the apparent resistivities appearing in Figure A-6 are shown in Figure A-7.

Appendix A: Additional Input Instructions

```
SCHLUMBERGER ELECTRIC SOUNDING

Apparent resistivities calculated using the linear filter method

electrode spacing   apparent resistivity
     1.00               1.21072
     1.47               1.51313
     2.15               2.07536
     3.16               2.95097
     4.64               4.19023
     6.81               5.87513
     10.0               8.08115
     14.7               10.8029
     21.5               13.8229
     31.6               16.5158
     46.4               17.7689
     68.1               16.4943
     100.               12.8532
     147.               8.79979
     215.               6.30746
     316.               5.40524
     464.               5.15234
     681.               5.06595
     1000.              5.02980
```

Figure A-6. A model-output file.

```
jif @
@electrode@
l1 [ar1]21:27
l1 [ar2]21:27
l1 [ar3]21:27
l1 [ar4]21:27
l1 [ar5]21:27
l1 [ar6]21:27
l1 [ar7]21:27
l1 [ar8]21:27
l1 [ar9]21:27
l1 [ar10]21:27
l1 [ar11]21:27
l1 [ar12]21:27
l1 [ar13]21:27
l1 [ar14]21:27
l1 [ar15]21:27
l1 [ar16]21:27
l1 [ar17]21:27
l1 [ar18]21:27
l1 [ar19]21:27
```

Figure A-7. A JUPITER instruction file.

   To navigate a single line of an output file, a number of instruction items can appear on a single line of an instruction file. The single line can be continued onto new lines using the continuation instructions. The items on the single or continued instruction line need to be separated from each other by at least one space. The start of a new instruction line signifies that the MIO_EXT subroutine of the Model Input-Output Module needs to read at least one new model-output file line. The number of lines read depends on the first instruction on the new instruction line.

MIO_EXT reads a model-output file in the forward (top-to-bottom) direction, using the instructions in the associated instruction file. Instructions should be written with this in mind; an instruction cannot direct MIO_EXT to "backtrack" to a previous line of the model-output file. Also, because model-output file lines are processed from left to right, an instruction cannot direct MIO_EXT backwards to an earlier part of a line.

Primary Marker

Unless it is a continuation of a previous line, each instruction line needs to begin with either of two instruction items: a primary marker or a line advance item (Table A-2). The primary marker has already been discussed briefly. It is a string of characters, bracketed at each end by a marker delimiter. If a marker is the first item on an instruction line, then it is a primary marker; if it occurs later in the line, following other instruction items, it is a secondary marker, the operation of which is discussed below after the sections "Line Advance" and "Continuations".

On encountering a primary marker in an instruction file, MIO_EXT reads the model-output file, line by line, searching for the string between the marker delimiter characters. When it finds the string, it places its "cursor" at the last character of the string. The cursor is never actually seen by the application user; it simply marks the point where MIO_EXT is located in its processing of the model-output file. If any further instructions on the same instruction line as the primary marker direct MIO_EXT to further processing of this line, that processing starts at the end of the primary marker.

If there are blank characters in a primary (or secondary) marker, exactly the same number of blank characters is expected in the matching string in the model-output file.

Often, as in Figure A-7, a primary marker is part or all of a header or label; such a header or label often precedes a model's listing of the outcomes of its calculations and thus makes a convenient reference point from which to search. Searching for a primary marker is a time-consuming process as each line of the model-output file is individually read and scanned for the marker. If the same dependents are always written to the same lines as counted from the top of a model-output file, it is most efficient to use the line advance item described in the "Line Advance" section.

A primary marker may be the only item on a JUPITER instruction line, or it may precede a number of other items directing further processing of the line containing the marker. In the former case the purpose of the primary marker is simply to establish a reference point for further downward movement within the model-output file as set out in subsequent instruction lines.

Primary markers can provide a useful means of navigating a model-output file. Consider the fragment from a model-output file shown in Figure A-8 (the dots replace one or a number of lines not shown in the figure to conserve space). The instruction file fragment shown in Figure A-9 provides a means to read the numbers comprising the third solution vector. The "SOLUTION VECTOR" primary marker is preceded by the primary marker "PERIOD NO. 3". The "PERIOD NO. 3" primary marker is used purely to establish a reference point from which a search can be made for the "SOLUTION VECTOR" marker; if this reference point were not established (using either a primary marker or line advance item), MIO_EXT would read the solution vector pertaining to a previous time period.

```
                   .
                   .
   TIME PERIOD NO. 1 --->
                   .
                   .
   SOLUTION VECTOR:
     1.43253   6.43235   7.44532   4.23443   91.3425   3.39872
                   .
                   .
   TIME PERIOD NO. 2 --->
                   .
   SOLUTION VECTOR
     1.34356   7.59892   8.54195   5.32094   80.9443   5.49399
                   .
   TIME PERIOD NO. 3 --->
                   .
                   .
   SOLUTION VECTOR
     2.09485   8.49021   9.39382   6.39920   79.9482   6.20983
```

Figure A-8. Fragment of a model-output file.

```
jif *
       .
       .
*PERIOD NO. 3*
*SOLUTION VECTOR*
l1 (obs1)5:10 (obs2)12:17 (obs3)21:28 (obs4)32:37 (obs5)41:45
& (obs6)50:55
       .
       .
```

Figure A-9. Fragment of an instruction file.

Line Advance

The syntax for the line advance item is "Ln" where n is the number of lines to advance; the L may be either uppercase or lowercase. The line advance item needs to be the first item of an instruction line. As previously explained, the initial item in an instruction line is always a directive to the MIO_EXT subroutine to move at least one line farther in its perusal of the model-output file (unless it is a continuation character). In the case of the primary marker, MIO_EXT stops reading new lines when it finds the pertinent text string. For a line advance, however, it does not need to examine model-output file lines as it advances. It simply moves forward n lines, placing its processing cursor just before the beginning of this new line, this point becoming the new reference point for further processing of the model-output file.

Normally a line advance item is followed on the same line by other instructions. However, if the line advance item is the only item on an instruction line, this does not break any syntax rules.

In Figure A-6, model-calculated apparent resistivities are written on subsequent lines. Hence, before reading each dependent, MIO_EXT is instructed to move to the beginning of a new line using the "l1" line advance item (see Figure A-9).

123

Appendix A: Additional Input Instructions

If a line advance item leads the first instruction line of a JUPITER instruction file, the reference point for line advance is taken as a "dummy" line just above the first line of the model-output file. Thus if the first instruction line begins with "l1", processing of the model-output file begins on its first line; similarly, if the first instruction line begins with "l8", processing of the model-output file begins at its eighth line.

Continuation

An instruction line can be broken between any two instructions by using the continuation character, "&", to inform the Model Input-Output Module that a certain instruction line is actually a continuation of the previous line. Thus the instruction file fragment

```
l1 %RESULTS% %TIME (4)% %=% !obs1! !obs2! !obs3!
```

is equivalent to

```
l1
& %RESULTS%
& %TIME (4)%
& %=%
& !obs1!
& !obs2!
& !obs3!
```

For both these fragments, the marker delimiter is assumed to be "%". Note that the continuation character must be separated from the instruction which follows it by at least one space.

Secondary Marker

A secondary marker does not occupy the first position of a JUPITER instruction line and, therefore, does not direct the MIO_EXT subroutine to move downward on the model-output file (though it can be instrumental in this – see below). A secondary marker initiates a search along the current model-output file line until the secondary marker string is found. The cursor is placed on the last character of the string, ready for subsequent processing of the line.

Figure A-10 shows a fragment of a model-output file; Figure A-11 shows the instructions needed to read the potassium (K) concentration. A primary marker is used to place the MIO_EXT cursor on the line above the line with results for the distance of interest. Then MIO_EXT is directed to advance one line and read the number following the "K:" string in order to find a dependent named "kc"; the exclamation marks surrounding "kc" are discussed in the Non-Fixed Extraction Instructions" section.

```
              .
              .
DISTANCE = 20.0: CATION CONCENTRATIONS:-
Na: 3.49868E-2  Mg: 5.987638E-2  K: 9.987362E-3
              .
              .
```

Figure A-10. Fragment of a model-output file.

```
jif ~
      .
      .
      .
~DISTANCE = 20.0~
ll ~K:~ !kc!
      .
      .
```

Figure A-11. Fragment of an instruction file.

A useful feature of the secondary marker is illustrated in Figure A-12 and A-13, which show a model-output file fragment and a corresponding instruction file fragment, respectively. If a secondary marker is preceded only by other markers (including, perhaps, one or more secondary markers and certainly a primary marker), and the text string corresponding to that secondary marker is not found on a model-output file line on which the previous markers' strings have been located, MIO_EXT assumes that it has not yet found the correct model output line and resumes its search for a line that holds the text from all three markers. In the example shown in Figure A-12, the instruction "%TIME STEP 10%" causes MIO_EXT to pause on its downward journey through the model-output file at the first line illustrated. When the string "STRAIN" is not found on the same line, it recommences its search, again looking for the string "TIME STEP 10". If a line is found containing both the primary and secondary markers, execution of subsequent instructions are executed. If no such line is found, execution is returned to the calling subprogram with an appropriate error message.

```
           .
           .
           .
TIME STEP 10 (13 ITERATIONS REQUIRED) STRESS --->
X = 1.05 STRESS = 4.35678E+03
X = 1.10 STRESS = 4.39532E+03
           .
           .
           .
TIME STEP 10 (BACK SUBSTITUTION) STRAIN --->
X = 1.05 STRAIN = 2.56785E-03
X = 1.10 STRAIN = 2.34564E-03
           .
           .
           .
```

Figure A-12. Fragment of a model-output file.

```
jif %
      .
      .
      .
%TIME STEP 10% %STRAIN%
ll %STRAIN =% !str1!
ll %STRAIN =% !str2!
      .
      .
```

Figure A-13. Fragment of an instruction file.

Appendix A: Additional Input Instructions

Whitespace

The whitespace instruction directs MIO_EXT to move the cursor forward from its current position until it encounters the next blank space. MIO_EXT then moves the cursor forward again until it finds a nonblank character. The cursor is placed just before the nonblank character (that is, on the last blank space in a sequence of blank characters) ready for the next instruction.

The whitespace instruction is a simple "W," separated from its neighbouring instructions by at least one blank space. The W may be either uppercase or lowercase.

Consider the model-output file line:

```
MODEL OUTPUTS:   2.89988   4.487892   -4.59098   8.394843
```

The following instruction line directs MIO_EXT to read the fourth number on the above line:

```
%MODEL OUTPUTS:% w w w w !obs1!
```

The instruction line begins with a primary marker. After this marker is found and processed, the MIO_EXT cursor rests on the ":" character of "OUTPUTS:", the last character of the marker string. In response to the first whitespace instruction, MIO_EXT finds the next whitespace and then moves the cursor to the end of this whitespace, that is, just before the "2" of the first number on the model-output file line. The second whitespace instruction moves the cursor to the blank character preceding the first "4" of the second number on the above line; processing of the third whitespace instruction results in MIO_EXT moving its cursor to the blank character just before the negative sign. After the fourth whitespace instruction, the cursor rests on the blank character preceding the last number; the latter is then read as a non-fixed dependent. Examples are presented below in the section "Example Instruction Files".

Tab

The tab instruction places the MIO_EXT cursor at a user-specified character position (that is, column number) on the model-output file line that it is currently processing. The instruction syntax is "Tn" where n is the column number; the T may be either uppercase or lowercase. The column number is obtained by counting positions (including blank spaces) from the left side of any line, starting at 1. Like the whitespace instruction, the tab instruction is used to navigate through a model-output file line prior to locating and reading a number.

For example, consider the line from a model-output file shown in Figure A-14. The instructions shown in Figure A-14 assume that the cursor was previously four lines above the line shown; the marker delimiter is defined to be "%." Implementation of the "t34" instruction places the cursor on the ":" following the "TIME(2)" string. The secondary search string then moves the cursor to the next "=" character. From there, the last number on the above line can be read as a non-fixed dependent.

Appendix A: Additional Input Instructions

Line from a model output file	. . . `TIME(1): A = 1.34564E-04, TIME(2): A = 1.45654E-04`
Column numbers	`1234567890123456789012345678901234567890123456789` `        10        20        30        40        50`
Instruction file	jif % l4 t34 %=% !a2!

Figure A-14. Line from a model-output file and an instruction file illustrating the use of the tab instruction. The number to be read is shaded. Dots replace one or more lines that are not shown.

Example Instruction Files

The examples are organized on the basis of their use of fixed, semi-fixed, and non-fixed extraction instructions. In the following discussion, dependents to be extracted using these types of instructions are referred to, respectively, as "fixed", "semi-fixed", and "non-fixed" dependents. To allow verification of results of extraction instructions, IVERB of the Global Data Module (Appendix C) can be set to a value greater than 3, causing MIO_EXT (Appendix F) to write the value extracted for each dependent to unit IOUT.

Fixed Extraction Instructions

Extraction using fixed extraction instructions is by far the most efficient way to read a dependent value because no searching is required; instead, a number is simply read from the space identified.

Figure A-15 shows how the numbers listed in the third solution vector of Figure A-8 can be read as fixed dependents. The instruction item informing MIO_EXT how to read a fixed dependent consists of two parts. The first part consists of the dependent's name enclosed in square brackets, while the second part consists of the first and last columns from which to read the dependent. No spaces are allowed; the Model Input-Output Module always construes a space in an instruction file as marking the end of one instruction item and the beginning of another, unless the space lies between marker delimiters.

```
jif *
.        .
*PERIOD NO. 3*
*SOLUTION VECTOR*
11 [obs1]1:9 [obs2]10:18 [obs3]19:27 [obs4]28:36 [obs5]37:45
& [obs6]46:54
     .
     .
```

Figure A-15. Fragment of an instruction file.

Reading numbers as fixed dependents is useful when the model writes its output in tabular form by using fixed-field-width specifiers. The range defined by the column numbers must be wide enough to accommodate the maximum length that the number will occupy in the course of all model runs. If it is not wide enough, MIO_EXT may read only a truncated part of the number or omit a negative sign preceding the number. However the range must not be so wide that it includes part of another number; in this case, a run-time error may occur. If the two numbers abut one another, an incorrect value may be read by MIO_EXT. If an error is encountered, subroutine MIO_EXT returns control to its calling program, with an appropriate error message assigned to the AMESSAGE string and the offending instruction stored in the INSTRUCTION string.

Where a model writes its results in the form of an array of numbers, it is not an uncommon occurrence for these numbers to abut each other. Consider, for example, the following Fortran code fragment:

```
      A=1236.567
      B=8495.0
      C=-900.0
      WRITE(10,20) A,B,C
20    FORMAT(3(F8.3))
```

The result will be

```
1236.5678495.000-900.000
```

In this case there is no choice but to read these numbers as fixed dependents. The two alternative extraction methods require that the number be surrounded by either whitespace or a string that is invariant from model run to model run and can thus be used as a marker. The above three numbers can be read as dependents A, B, and C by using the following instruction line:

```
l1 [A]1:8 [B]9:16 [C]17:24
```

If an instruction line contains only fixed dependents, there is no need for it to contain any whitespace or tabs. The only reason to use a secondary marker is in conjunction with a primary marker to determine which model-output file line the MIO_EXT cursor should settle on, as discussed previously. These instructions normally are used to navigate through a model-output file line prior to reading a non-fixed dependent; such navigation is not required to locate a fixed dependent as its location on a model-output file line is defined without ambiguity by the column numbers included within the fixed extraction instruction.

Semi-Fixed Extraction Instructions

Figure A-9 demonstrates the use of semi-fixed dependents. Semi-fixed dependents are similar to fixed dependents in that two numbers are provided to locate the dependent's position by column number on the model-output file. However, in contrast to fixed dependents, these numbers do not locate the dependent exactly. When MIO_EXT encounters a semi-fixed extraction instruction, it proceeds to the first of the two nominated column numbers and then, if this column is occupied by a blank character, it searches the output file line from left to right beginning at this column number, until it reaches either the second identified column or a non-blank character. If it reaches the second column before finding a non-blank character, an error condition arises. If it finds a non-blank character, it then locates the nearest whitespace on either side of the character; in this way, it identifies one or a number of non-blank characters sandwiched between whitespace ("whitespace" includes the beginning and (or) the end of the model-output file line). It tries to read these characters as a number, and associate it with the dependent named in the semi-fixed extraction instruction. The width of this number can be greater than the range defined by the column numbers cited in the semi-fixed extraction instruction.

For semi-fixed dependents, the dependent name is enclosed in round brackets rather than square brackets (Table A-2). Again, there must be no space separating the two parts of the semi-fixed extraction instruction. The column numbers must be in ascending order.

Reading a number as a semi-fixed dependent is useful if it is unclear how large the representation of the number could stretch on a model-output file as its magnitude grows or diminishes in different model runs. It is useful to know whether the number is left or right justified. It must be ensured that at least part of the number will always fall between (and including) the two nominated columns and that, whenever the number is written and whatever its size, it will always be surrounded either by whitespace or by the beginning or end of the model-output file line. If MIO_EXT encounters only whitespace between (and including) the two nominated column numbers, or if it encounters non-numeric characters or two number fragments separated by whitespace, an error condition occurs and MIO_EXT returns control to its calling program unit after assigning an appropriate error message to the AMESSAGE string.

As for fixed dependents, it is normally not necessary to have secondary markers, whitespace, and tabs on the same line as a semi-fixed dependent because the column numbers provided with the semi-fixed extraction instruction determine the location of the dependent on the line. As always, dependents must be read from left to right on any one instruction line; if more than one semi-fixed extraction instruction is provided on a single instruction line, the column numbers must increase from left to right.

For the case illustrated in Figures A-6 and A-7, all the fixed dependents could have been read as semi-fixed dependents, with the difference between the column numbers either remaining the same or being reduced to substantially smaller than those shown in Figure A-7. However, it takes more effort for subroutine MIO_EXT to read a semi-fixed dependent than it does for it to read a fixed dependent as it must establish for itself the extent of the number to be read.

After MIO_EXT reads a semi-fixed dependent, its cursor resides at the end of the number which it has just read. Any further processing of the line must take place to the right of that position.

Non-Fixed Extraction Instructions

Figures A-11 and A-13 demonstrate the use of non-fixed dependents. A non-fixed extraction instruction does not include any column numbers — the number to be read is found using secondary markers, whitespace, or tabs that precede the non-fixed dependent on the instruction line.

If it is not known exactly where, on a particular model-output file line, a number is to be written, but the structure of the line is known, then this knowledge can be used to find the number. In the instruction file, a non-fixed dependent is represented simply by the name of the dependent surrounded by exclamation marks; as usual, no spaces can separate the exclamation marks from the dependent name as MIO_EXT interprets spaces in an instruction file as denoting the end of one instruction item and the beginning of another.

When the MIO_EXT subroutine encounters a non-fixed extraction instruction, it first searches forward from its current cursor position until it finds a non-blank character; MIO_EXT assumes this character is the beginning of the number representing the non-fixed dependent. Then it searches forward again until it finds either a blank character, the end of the line, or the first character of a secondary marker that follows the non-fixed extraction instruction in the instruction file; MIO_EXT assumes that the number representing the non-fixed dependent finishes at the previous character position. In this way it identifies a string of characters that it tries to read as a number; if it is unsuccessful in reading a number because of the presence of non-numeric characters or some other problem, MIO_EXT returns execution to its calling program, leaving an appropriate error message in the AMESSAGE string. An error condition also arises if MIO_EXT encounters the end of a line while looking for the beginning of a non-fixed dependent.

Consider the output file fragment shown in Figure A-16. The species populations at different times cannot be read as either fixed or semi-fixed dependents. The numbers representing these populations cannot be guaranteed to fall within a certain range of column numbers on the model-output file, because "iterative adjustment" may be required in the calculation of any such population. Hence MIO_EXT must find its way to the number using another method; one such method is illustrated in Figure A-17.

```
        .
        .
    SPECIES POPULATION AFTER 1 YEAR  = 1.23498E5
    SPECIES POPULATION AFTER 2 YEARS = 1.58374E5
    SPECIES POPULATION AFTER 3 YEARS (ITERATIVE ADJUSTMENT REQUIRED)= 1.78434E5
    SPECIES POPULATION AFTER 4 YEARS = 2.34563E5
        .
        .
```

Figure A-16. Fragment of a model-output file.

```
jif *
        .
        .
*SPECIES* *=* !sp1!
l1 *=* !sp2!
l1 *=* !sp3!
l1 *=* !sp4!
        .
        .
```

Figure A-17. Fragment of a JUPITER instruction file.

A primary marker is used to move the MIO_EXT cursor to the first of the lines shown in **Figure A-16**. Then, noting that the number representing the species population always follows a "=" character, the "=" character is used as a secondary marker. After it processes a secondary marker, the MIO_EXT cursor always resides on the last character of the marker; in this case on the "=" character itself. After reading the "=" character, MIO_EXT can process the instruction !sp1! by isolating the string "1.23498E5" in the manner described previously.

After it reads the model-calculated value for dependent "sp1," MIO_EXT moves to the next instruction line. In accordance with the "l1" instruction, the next line of the model-output file is read into memory. MIO_EXT then searches for a "=" character and reads the number following this character as dependent "sp2." This procedure is then repeated for dependents "sp3" and "sp4."

Successful performance of a non-fixed extraction instruction depends on the instructions preceding it. The secondary marker, tab, and whitespace instructions are most useful in this regard, though fixed and semi-fixed dependents may also precede a non-fixed dependent; remember that in all these cases MIO_EXT places its cursor over the last character of the string or number it identifies on the model-output file corresponding to an instruction item, before proceeding to the next instruction.

Appendix A: Additional Input Instructions

Consider the model-output file line shown below as a further illustration of the use of non-fixed dependents.

```
4.33 -20.3 23.392093 3.394382
```

To extract the fourth number when it is not known whether the numbers preceding it may be written with greater precision in some model runs (hence pushing the number of interest to the right), a non-fixed dependent and whitespace can be used as follows:

```
l10 w w w !obs1!
```

Here it is assumed that, prior to reading this instruction, the MIO_EXT cursor was located on the 10th preceding line of the model-output file. As long as no whitespace ever precedes the first number, this instruction line performs correctly. However, if whitespace sometimes precedes the first number, then the first number can be read as a dummy dependent as:

```
l10 !dum! w w w !obs1!
```

As was explained previously, the number corresponding to a dependent named "dum" is not actually used; nor can the name "dum" be used as an observation or prediction name. The use of "dum" is reserved for instances like this, where a number must be read to navigate along a line of the model-output file. The number is read according to the non-fixed extraction protocol, for only dependents of this type can be dummy dependents.

An alternative to the use of whitespace in locating the "obs1" dependent in the above example could involve using the dummy dependent more than once. Hence, the instruction line below would also enable the number representing "obs1" to be located and read:

```
l10 !dum! !dum! !dum! !obs1!
```

If the numbers in the above example were separated by commas instead of whitespace, the commas could be used as secondary markers.

A number not surrounded by whitespace can be read as a non-fixed dependent with the proper choice of secondary markers. Consider the model-output file line shown below:

```
SOIL WATER CONTENT (NO CORRECTION)=21.345634%
```

It may not be possible to read the soil water content as a fixed dependent because the "(NO CORRECTION)" string may or may not be present after any particular model run. Reading it as a non-fixed dependent appears troublesome as the number is neither preceded nor followed by whitespace. However, a suitable instruction line is

```
l5 *=* !sws! *%*
```

Here the secondary marker "%" is referenced after the dependent being read. If this marker were absent, a run-time error would occur when MIO_EXT tried to read the soil water content, because it would define the dependent string to include the non-numeric "%" character. By including the "%" character as a secondary marker after the number representing the "sws" dependent, MIO_EXT separates the character from the string before trying to read the number. If a post-dependent secondary marker of this type begins with a numerical character, MIO_EXT may include this character with the dependent number if there is no whitespace separating it from the dependent.

The fact that there is no whitespace between the "=" character and the number that must be read causes no problems either. After processing of the "=" character as a secondary marker, the MIO_EXT processing cursor falls on the "=" character itself. The search for the first non-blank character initiated by the instruction !sws! terminates on the very next character after the "="; that is, the "2" character. MIO_EXT then accepts this character as the left boundary of the string from which it must read the soil moisture content and searches forward for the right boundary of the string in the usual manner.

After MIO_EXT reads a non-fixed dependent, it places its cursor on the last character of the dependent number. It can then undertake further processing of the model-output file line to read further non-fixed, fixed, or semi-fixed dependents or process navigational instructions as directed.

Making an Instruction File

An instruction file is easily built using a text editor. Normally two windows should be open — one to write the instructions and the other to view the model-output file, which is the focus of the instructions.

Caution must always be exercised in building an instruction set, especially if navigational instructions such as markers, whitespace, tabs, and dummy dependents are used. The Model Input-Output Module will always follow user-supplied instructions to the letter, but it may not read the numbers that the user intends. If MIO_EXT tries to read a dependent but does not find a number where it expects to find one, a run-time error occurs. MIO_EXT reports the error, including the instruction it was implementing when the error condition arose. The error message reported by the JUPITER application allows the problem to be identified. A wrong number read by MIO_EXT from the model-output file, however, may only become apparent if the JUPITER application appears to malfunction as a result. For example, if a parameter estimation process reports an unusually high objective function or the value of the objective function cannot be lowered. In this case, execution of the JUPITER application should be stopped, the misread number found, and the instruction file corrected.

Derivatives Interface File

A Derivatives Interface File provides a JUPITER application with information needed to obtain model-calculated sensitivities (derivatives of dependents with respect to parameters) from a model-output file. All items are read in free format. The format for a Derivatives Interface File is shown in Table A-3.

Appendix A: Additional Input Instructions

Table A-3. Derivatives Interface File format.

Item	Variable name or literal contents	Explanation
0	# Text	Zero or more comment lines allowed only at the top of the file. Comments are identified by # in column 1. Comments can be placed to the right of the values to be read, except not on lines containing names of parameters or dependents. See example Derivatives Interface file in Figure A-18.
1	DERFILE	Name of the model-generated file containing derivatives.
2	NSKIP	Number of lines at the top of the file DERFILE to skip before reading derivative values.
3	NDEP NPAR	Number of dependents; number of parameters.
4	ORIENTATION	Either "**ROW/DEP**" or "**ROW/PAR**". Enter with or without quotes. See note 1 below.
5	DERFORMAT	Fortran format for reading derivatives values, or: "**(FREE)**". See note 2 below.
6	"PARAMETERS"	Enter the word "PARAMETERS", with or without quotation marks. Interpretation is case insensitive.
7	Parameter names	NPAR parameter names. The names correspond, in order, to the parameters for which model-calculated derivatives are provided in file DERFILE. See note 3 below.
8	"DEPENDENTS"	Enter the word "DEPENDENTS", with or without quotation marks. Interpretation is case insensitive.
9	Dependent names	NDEP dependent names. The names correspond, in order, to the dependents for which model-calculated derivatives are provided in file DERFILE. See note 3 below.

Notes:
1. ORIENTATION: The choice supports reading either an untransposed or a transposed jacobian matrix from the model-generated file. "ROW/DEP" would indicate that each row in file DERFILE contains derivatives for one dependent. "ROW/PAR" would indicate that each row contains derivatives for one parameter.
2. DERFORMAT: The format string needs to include parentheses. Single or double quotes may be used to include embedded spaces or commas. Length limit: 200 characters.
3. Parameter names and dependent names: Names are read until NPAR (or NDEP) names have been read; the names should be in the order used in the model-generated derivatives file. Multiple names may be listed on each line. The names need to correspond to parameter or dependent names defined elsewhere in program input.

An example Derivatives Interface File is shown in Figure A-18.

```
# Derivatives Interface File for tc1 model
tc1._su          (DERFILE)
1                (NSKIP)
35  9            (NDEP NPAR)
row/dep          (ORIENTATION)
'(20x,9f15.0)'   (DERFORMAT)
PARAMETERS
wells_tr  rch_zone_1  rch_zone_2  rivers  ss_1  hk_1  vert_k_cb  ss_2  hk_2
DEPENDENTS
h1.0  h1.1   h1.12  h2.0   h2.1   h2.2   h2.8   h2.12  h3.0  h3.1  h3.12  h4.0
h4.1  h4.12  h5.0   h5.1   h5.12  h6.0   h6.1   h6.12  h7.0  h7.1  h7.12  h8.0
h8.1  h8.12  h9.0   h9.1   h9.12  h0.0   h0.1   h0.12  SS    TR3   TR12
```

Figure A- 18. Example Derivatives Interface File, with an intial comment line and comments to the right of the values to be read on the subsequent five lines.

Matrix Input and Output

Input and output formats for matrices are described in this section. The Complete Matrix format is used when the matrix may be fully populated or otherwise not suitable for compression. The Compressed Matrix format saves storage space when the matrix contains a large number of zero values. Each matrix can be associated with a name, optionally provided on the first line, following the keyword identifying the format type. In both cases CONTROLRECORD is an optional keyword. As for the input blocks described in Chapter 3, the required keywords are shown here in bold capital letters. Input items listed in square brackets are optional.

Complete Matrix

The complete matrix input format has one of the following two forms, where "COMPLETEMATRIX" is a keyword (case-insensitive), NR is the number of rows in the matrix, and NC is the number of columns in the matrix:

```
COMPLETEMATRIX   [NAME]
NR   NC
2-D array
```

or

```
COMPLETEMATRIX   [NAME]
NR   NC   CONTROLRECORD
array-control record
[2-D array]
```

If the optional keyword CONTROLRECORD is absent, the first form is used, and the array data, readable in free format, are listed after the line containing the matrix dimensions NR and NC.

134

Appendix A: Additional Input Instructions

If the keyword CONTROLRECORD is present, the second form is used, in which an array-control record follows the line containing the matrix dimensions NR and NC. The array-control record may take any of the following three forms:

```
CONSTANT   CNSTNT
INTERNAL   CNSTNT   FMTIN   IPRN
OPEN/CLOSE   FNAME   CNSTNT   FMTIN   IPRN
```

where:
CONSTANT, INTERNAL, and OPEN/CLOSE are keywords,
CNSTNT is a floating-point number or an integer,
FMTIN is a character string indicating the format in which values are to be read, and
IPRN is a print code.

One of the keywords CONSTANT, INTERNAL, or OPEN/CLOSE needs to be the first item in the array-control record, in any combination of uppercase and lowercase letters.

If the keyword is CONSTANT, the entire array is populated with the value supplied for CNSTNT.

If the keyword is INTERNAL, the array is populated with values read from unit INUNIT in the format indicated by FMTIN, and multiplied by CNSTNT.

If the keyword is OPEN/CLOSE, the file named FNAME is opened, and the array is populated with values read from file FNAME, read in the format indicated by FMTIN, and multiplied by CNSTNT. File FNAME is then closed.

The interpretation of IPRN depends on the type of the array being populated, as shown in Table A- 4.

Appendix A: Additional Input Instructions

Table A- 4. Supported values of IPRN and the associated formats.

[The meaning of IPRN depends on the type of BUF. The format associated with IPRN=0 is used when IPRN exceeds the largest supported value or is less than 0.]

IPRN	BUF is DOUBLE PRECISION or REAL	BUF is INTEGER
0	10G11.4	10I11
1	11G10.3	60I1
2	9G13.6	40I2
3	15F7.1	30I3
4	15F7.2	25I4
5	15F7.3	20I5
6	15F7.4	10I11
7	20F5.0	25I2
8	20F5.1	15I4
9	20F5.2	10I6
10	20F5.3	Not supported
11	20F5.4	Not supported
12	10G11.4	Not supported
13	10F6.0	Not supported
14	10F6.1	Not supported
15	10F6.2	Not supported
16	10F6.3	Not supported
17	10F6.4	Not supported
18	10F6.5	Not supported
19	5G12.5	Not supported
20	6G11.4	Not supported
21	7G9.2	Not supported

Appendix A: Additional Input Instructions

Compressed Matrix

The compressed matrix input format has one of the two following forms:

```
COMPRESSEDMATRIX   [NAME]
NNZ  NR  NC
IPOS(1)    VAL(1)
IPOS(2)    VAL(2)
...
IPOS(NNZ)  VAL(NNZ)
```

or

```
COMPRESSEDMATRIX   [NAME]
NNZ  NR  NC  [CONTROLRECORD]
array-control record
[IPOS(1)    VAL(1)
 IPOS(2)    VAL(2)
 ...
 IPOS(NNZ)  VAL(NNZ)]
```

where,

COMPRESSEDMATRIX is a keyword (case-insensitive),

NNZ is the number of non-zero values in a matrix assumed to have dimensions (NR, NC),

NR is the number of rows,

NC is the number of columns in the matrix.

IPOS(i) is the position in the true matrix of the ith non-zero entry in the matrix, assuming column-major storage order (In column-major storage order, all entries of column 1 are listed first, starting at row 1, followed by all entries of column 2, and so on.),

VAL(i) is the corresponding non-zero value.

CONTROLRECORD is an optional keyword.

If the **CONTROLRECORD** keyword is absent, the first form shown above is used and the non-zero elements of the array are in columns of IPOS and VAL values, readable in free format.

If the keyword **CONTROLRECORD** is present, the second form is used, in which the third line is an array-control record. The array-control record may take either of the following two forms:

```
INTERNAL   CNSTNT  FMTIN  IPRN
OPEN/CLOSE  FNAME  CNSTNT  FMTIN  IPRN
```

where:

INTERNAL and **OPEN/CLOSE** are keywords,

As defined previously in this section:

CNSTNT is a floating-point number or an integer,

FMTIN is a character string indicating the format in which values are to be read, and

IPRN is a print code.

Appendix A: Additional Input Instructions

One of the keywords INTERNAL or OPEN/CLOSE must be the first token on the array-control record, in any combination of upper- and lowercase letters.

If the keyword is INTERNAL, the array is populated with values read from unit INUNIT in the format indicated by FMTIN, and multiplied by CNSTNT.

If the keyword is OPEN/CLOSE, the file named FNAME is opened, and the array is populated with values read from file FNAME read in the format indicated by FMTIN, and multiplied by CNSTNT. File FNAME is then closed.

The interpretation of IPRN depends on the type of the array being populated, as shown in Table A- 4.

Appendix B: Datatypes (TYP) Module of Chapter 4

Public Generic Interfaces

The Datatypes Module contains two generic interfaces developed to nullify pointers and deallocate allocatable arrays in structures of derived data types. The derived data types are described in Chapter 4.

Interface TYP_DEALLOC

Description: This generic subroutine interface deallocates all allocatable arrays and nullifies all pointers in a derived type structure. This interface should not be used with structures that contain pointers for which the allocation status is undefined. Structures that contain pointers for which the allocation status is undefined (as is the case when the structure is declared) can be initialized by the TYP_NULL interface.

Argument list:
(DTS)

Declarations for argument-list variables:
[various types], INTENT(INOUT) :: DTS

Explanation of arguments:
DTS is a structure of a derived data type. DTS may be of type CDMATRIX, DMATRIX, or CHAR_1D_30. DTS also may be a pointer to a structure of type LNODE or LLIST.

Interface TYP_NULL

Description: This generic subroutine interface operates on a structure of a derived data type, nullifying any and all components of the structure that have the pointer attribute. It is good practice to call TYP_NULL as soon as a structure of a derived data type is declared. To avoid "memory leak," this interface should not be used on structures in which component arrays are allocated. The TYP_DEALLOCATE interface can be used to deallocate allocatable-array components of derived-type structures.

Argument list:
(DTS)

Declarations for argument-list variables:
TYPE (*any-type*), INTENT(INOUT) :: DTS

Explanation of arguments:
DTS is a structure of derived data type *any-type*, where *any-type* is one of: LNODE, LLIST, DMATRIX, CDMATRIX, or CHAR_1D_30.

Appendix C: Global Data (GLO) Module of Chapter 5

Public Data

Declaration of variables:
```
CHARACTER(LEN=14),   PARAMETER :: VERSIONID = '1.0.0'
REAL                           :: REALLOCAL
DOUBLE PRECISION               :: DOUBLELOCAL
INTEGER                        :: INTEGERLOCAL
REAL,                PARAMETER :: BIGREAL = HUGE(REALLOCAL)
DOUBLE PRECISION,    PARAMETER :: BIGDOUBLE = HUGE(DOUBLELOCAL)
INTEGER,             PARAMETER :: BIGINTEGER = HUGE(INTEGERLOCAL)
INTEGER,             PARAMETER :: LENDNAM = 20
INTEGER,             PARAMETER :: MAXRECLDEFAULT = 2000
INTEGER,             PARAMETER :: MAX_STRING_LEN = 2000
INTEGER,             PARAMETER :: MAXVERB = 5
INTEGER,             PARAMETER :: MINVERB = 0
INTEGER,             PARAMETER :: NPPREC = 500
CHARACTER(LEN=100),  PARAMETER :: BLANKS = &
'                                                    &
&                                                    '
CHARACTER(LEN=100),  PARAMETER :: HYPHENS = &
'----------------------------------------------------&
&----------------------------------------------------'
CHARACTER(LEN=MAX_STRING_LEN) :: AMESSAGE = ' '
CHARACTER(LEN=80)             :: ERRSUB = ' '
INTEGER                       :: IVERB = 3
```

Explanation of variables:

VERSIONID is the version number of the JUPITER API; its value will be updated with each release. The string will always be in the form: "integer.integer.integer", where the first integer is the major release number, the second integer is the minor release number, and the third integer is the subrelease number. Each integer may be represented by one to four digits.

REALLOCAL is used with the HUGE function to assign a value to BIGREAL.

DOUBLELOCAL is used with the HUGE function to assign a value to BIGDOUBLE.

INTEGERLOCAL is used with the HUGE function to assign a value to BIGINTEGER.

BIGREAL is used as the default value for selected real variables.

BIGDOUBLE is used as the default value for selected double-precision variables.

BIGINTEGER is used as the default value for selected integer variables.

LENDNAM is the number of characters allowed in the name of a dependent (observation or prediction).

MAXRECLDEFAULT is used as a default value for the RECL specifier in selected OPEN statements.

MAX_STRING_LEN is the maximum length of a string allowed in many contexts. In particular, it is the maximum allowable length for a string in an input block.

MAXVERB is the largest defined value for IVERB.

MINVERB is the smallest defined value for IVERB.

NPPREC is the maximum number of parameters per record written to data-exchange files.

BLANKS is a string of 100 blank characters, to be used as needed for output.

HYPHENS is a string of 100 hyphen characters, to be used as needed for output.

AMESSAGE is used to store an error message when an error is encountered.

ERRSUB is a character string for an error header. Generally, this is the name of a program unit.

IVERB is used to control the verbosity of program output. The meanings are:

      0 - No extraneous output;

      1 - Write warnings only;

      2 - Write warnings and notes;

      3 - Write warnings and notes, and echo selected input (default);

      4 - Write warnings and notes, and echo all input; and

      5 - Write warnings, notes, echoed input, and miscellaneous information.

Appendix D: Utilities (UTL) Module of Chapter 6

Private Data

Declaration of variables:
```
INTEGER, PARAMETER                  :: MAXDX = 500
INTEGER                             :: KDXDEF = 0
CHARACTER(LEN=10), DIMENSION(MAXDX) :: DXEXT = ' '
INTEGER, DIMENSION(MAXDX)           :: IUDX = 0
```

Explanation of variables:

MAXDX is the maximum allowed number of data-exchange file types used by an application. This number includes explicitly and implicitly supported file types (Chapter 17).

KDXDEF is the number of data-exchange file types used by an application during a particular run of the application. KDXDEF is incremented at run time each time UTL_DX_OPEN is called with a new (previously unreferenced) extension.

DXEXT stores extensions of data-exchange files used by an application.

IUDX stores unit numbers for open data-exchange files.

Notes:

1. Each time UTL_DX_OPEN is called with a new (previously unreferenced) extension, an element of DXEXT is populated with that extension. The storage order of extensions in DXEXT is determined by the order in which previously unreferenced extensions are found in calls to UTL_DX_OPEN.

2. When UTL_DX_OPEN is called, the unit number on which the file is opened is stored in the element of IUDX corresponding to the element in DXEXT used for the file extension. When a data-exchange file with the specified extension is closed by UTL_DX_CLOSE, the corresponding element of IUDX is changed to 0.

Data-Exchange File Management Functions

This section documents three functions used to manage data-exchange files. The names are:
Function UTL_DX _CLOSE
Function UTL_DX _GETUNIT
Function UTL_DX _OPEN

Function UTL_DX _CLOSE

Description: This function closes the data-exchange file associated with argument EXT and returns 0. It sets the corresponding element of the IUDX array to 0.

Argument list:
```
(EXT, FILESTAT)
```

Result:
```
IUDATEX
```

Declarations for argument-list and result variables:
```
CHARACTER(LEN=*),                 INTENT(IN) :: EXT
CHARACTER(LEN=*), OPTIONAL, INTENT(IN) :: FILESTAT
INTEGER                                   :: IUDATEX
```

Explanation of arguments:
> EXT is a file extension previously used in a call to UTL_DX_OPEN. If the extension has not been used in a call to UTL_DX_OPEN, a warning is issued. The comparison with previously used extensions is case-insensitive.
> FILESTAT, if present, is one of the legal arguments for the STATUS= specifier in the Fortran CLOSE statement and has the same meaning. FILESTAT must evaluate to either "KEEP" or "DELETE". The default is "KEEP".

Function UTL_DX _GETUNIT

Description: This function returns the unit number associated with the data-exchange file opened with the extension identified by the argument EXT.
Argument list variable:
```
(EXT)
```

Result:
```
IUDATEX
```

Declarations for argument-list and result variables:
```
CHARACTER(LEN=*), INTENT(IN) :: EXT
INTEGER                      :: IUDATEX
```

Explanation of argument and result:
> EXT is an extension previously used to open a data-exchange file by a call to UTL_DX_OPEN.
> IUDATEX is the unit number associated with the data-exchange file of the type specified by EXT. If no file of that type is open, IUDATEX is returned as 0. If EXT has not been used in a call to UTL_DX_OPEN, a warning is written to the screen and IUDATEX is returned as –1.

Function UTL_DX _OPEN

Description: This function opens a data-exchange file with an extension identified by the EXT argument and returns the unit number associated with it. The name of the file is generated using FBASE as the file name base, a dot, and EXT as the extension.
Argument list:
```
(FBASE, EXT, FILESTAT, MAXRECL)
```

Result:
```
IUDATEX
```

Declarations for argument-list and result variables:
```
CHARACTER(LEN=*),                 INTENT(IN) :: FBASE
CHARACTER(LEN=*),                 INTENT(IN) :: EXT
CHARACTER(LEN=*), OPTIONAL, INTENT(IN) :: FILESTAT
INTEGER,          OPTIONAL, INTENT(IN) :: MAXRECL
INTEGER                                   :: IUDATEX
```

Explanation of arguments and result:

FBASE is the file-name base to be used in constructing the name of the data-exchange file.

EXT is the extension to be used in constructing the name of the data-exchange file; it also identifies the type of data-exhange file.

FILESTAT, if present, is one of the legal arguments for the STATUS= specifier in the Fortran OPEN statement and has the same meaning. FILESTAT must evaluate to either "UNKNOWN", "REPLACE", "SCRATCH", "OLD", or "NEW". The default is "UNKNOWN".

MAXRECL, if specified, overrides the default maximum record length, which may depend on the file type. For implicitly supported file types, the default maximum record length is MAXRECLDEFAULT of the Global Data Module (Appendix C).

IUDATEX is the unit number associated with the opened file. If EXT is blank, IUDATEX is returned as –1; if a problem opening the file is encountered, IUDATEX is returned as –2.

Notes:
1. Explicitly supported data-exchange file types are listed in Table 17-1. If EXT is not listed in Table 17-1, the file type and extension are implicitly supported.
2. Explicitly supported file types are opened with appropriate values for the RECL specifier of the OPEN statement. The RECL value for some file types depends on the value of NPPREC of the Global Data Module (Appendix C).
3. If on invocation of UTL_DX_OPEN a data-exchange file of the type identified by EXT is open, that file is closed and a new file is opened.

Data Input Subroutines, Functions, and Generic Interfaces

This section documents subroutines, functions, and generic interfaces used primarily to read input. The subroutines, functions, and interfaces are arranged alphabetically as follows.

Subroutine UTL_DX_READ_MCMV
Subroutine UTL_DX_READ_WT
Function UTL_GETARG
Function UTL_GETLINE
Interface UTL_READ2D
Subroutine UTL_READBLOCK
Interface UTL_READMATRIX

Subroutine UTL_DX_READ_MCMV

Description: This subroutine reads a data-exchange file of type "_mc" or "_mv". If the file to be read has been prepared correctly, the subroutine reads either a parameter correlation (_mc) or variance-covariance (_mv) matrix.

Argument list:
```
(EXT, NPE, OUTNAM, PARNAM, CMAT)
```

Declarations for argument-list variables:
```
CHARACTER(LEN=*),                      INTENT(IN)  :: EXT
INTEGER,                               INTENT(IN)  :: NPE
CHARACTER(LEN=*),                      INTENT(IN)  :: OUTNAM
CHARACTER(LEN=12), DIMENSION(NPE),     INTENT(OUT) :: PARNAM
DOUBLE PRECISION, DIMENSION(NPE,NPE),  INTENT(OUT) :: CMAT
```

Explanation of arguments:

EXT is the extension associated with the file type to be read. It must be either "_mc" or "_mv".

NPE is the number of adjustable parameters.

OUTNAM is the file-name base to be used in constructing the name of the data-exchange file to be read.

PARNAM is populated with parameter names read from the data-exchange file.

CMAT is populated with the matrix read from the data-exchange file. If EXT is "_mc", CMAT is populated with a parameter correlation matrix; if EXT is "_mv", CMAT is populated with a variance-covariance matrix.

Subroutine UTL_DX_READ_WT

Description: This subroutine reads a data-exchange "_wt" or "_wtpri" file (Table 17-3) and populates one or two CDMATRIX structures (Chapter 1) with data read from the file.

Argument list:
```
(IOUT, OUTNAM, EXT, WTMAT, WTMATSQR)
```

Declarations for argument-list and result variables:
```
INTEGER,                         INTENT(IN)    :: IOUT
CHARACTER(LEN=*),                INTENT(IN)    :: OUTNAM
CHARACTER(LEN=*),                INTENT(IN)    :: EXT
TYPE(CDMATRIX),                  INTENT(INOUT) :: WTMAT
TYPE(CDMATRIX), OPTIONAL,        INTENT(INOUT) :: WTMATSQR
```

Explanation of arguments:

IOUT is the unit number associated with the file to which error messages are to be written, if required.

OUTNAM is the root file name for output files.

EXT is one of two supported data-exchange file types to be created; its value may be either "_wt" or "_wtpri." If EXT is "_wt", matrix(ces) for dependents is (are) read from a data-exchange file of type _wt. If EXT is "_wtpri," matrix(ces) for prior information is (are) read from a data-exchange file of type _wtpri (Table 17-3).

WTMAT is populated with the first matrix in the "_wt" or "_wtpri" file with the base name OUTNAM.

WTMATSQR is populated with the second matrix in the "_wt" or "_wtpri" file with the base name OUTNAM, if WTMATSQR is present in the argument list and if the file contains two matrices.

Function UTL_GETARG

Description: This function returns an argument on the command line following the command name, up to MAX_STRING_LEN characters, where MAX_STRING_LEN is defined in the Global Data Module.

Argument list:
```
(NARG)
```

Result:
```
ARG
```

Declarations for argument-list and result variables:
```
INTEGER, INTENT(IN), OPTIONAL :: NARG
CHARACTER(LEN=MAX_STRING_LEN) :: ARG
```

Explanation of arguments:
> NARG, if present, is the number of the argument to be returned. The default is 1, indicating the first argument after the command name is to be returned.
> ARG is the argument in position number NARG, where position 1 is the first argument after the command name.

Notes:
> 1. The Fortran 90 standard (American National Statndards Instutute, Inc., 1992) does not support command line arguments. This subroutine makes use of a compiler-specific extension to Fortran 90 to obtain argument(s) from command line. Editing this subroutine may be required if the extension used in this subroutine is not supported by your compiler.
> 2. If single (') or double (") quote characters appear in the text following the command name in the command line, they are interpreted as delimiters for the command argument(s), allowing spaces or tab characters to be embedded in an argument.

Function UTL_GETLINE

Description: This function reads (starting from the current position in the file) one or more lines from unit INUNIT and returns the next non-blank line that is not a comment line, indicated by a '#' character in column 1. If the end of the file connected to unit INUNIT is reached without finding a non-blank, non-comment line, a blank line is returned.

Argument list:
```
(INUNIT, ICOM)
```

Result:
```
LINE
```

Declarations for argument-list and result variables:
```
INTEGER, INTENT(IN)            :: INUNIT
INTEGER, INTENT(IN), OPTIONAL :: ICOM
CHARACTER(LEN=MAX_STRING_LEN) :: LINE
```

Explanation of arguments:
> INUNIT is the unit number of the file from which lines are to be read.
> ICOM is an optional argument; if ICOM is present and greater than zero and if IVERB of the Global Data Module is greater than 3, any comment lines read from unit INUNIT are written to unit ICOM.

Interface UTL_READ2D

Description: This generic subroutine interface populates a 2-D array with data read from a specified file unit. The subroutine first reads an array-control record to establish how the data are to be read. The array-control record may take any of the following three forms:
```
CONSTANT   CNSTNT
INTERNAL   CNSTNT  FMTIN  IPRN
OPEN/CLOSE  FNAME  CNSTNT  FMTIN  IPRN
```

Where: CONSTANT, INTERNAL, and OPEN/CLOSE are keywords, CNSTNT is a floating-point number or an integer, FMTIN is a character string indicating the format in which values are to be read, and IPRN is a print code. One of the keywords CONSTANT, INTERNAL, or OPEN/CLOSE must be the first token on the array-control record, in any combination of upper- and lowercase letters. If the keyword is CONSTANT, the entire array is populated with the value supplied for CNSTNT.

If the keyword is INTERNAL, the array is populated with values read from unit INUNIT in the format indicated by FMTIN, and multiplied by CNSTNT.

If the keyword is OPEN/CLOSE, the file named FNAME is opened, and the array is populated with values read from file FNAME read in the format indicated by FMTIN, and multiplied by CNSTNT. File FNAME is then closed.

The interpretation of IPRN depends on the type of the array being populated, as described below.

Argument list:
```
(ANAME, NR, NC, INUNIT, IOUT, BUF, IPRN)
```

Declarations for argument-list variables:
```
CHARACTER(LEN=*),                      INTENT(IN)  :: ANAME
INTEGER,                               INTENT(IN)  :: NR
INTEGER,                               INTENT(IN)  :: NC
INTEGER,                               INTENT(IN)  :: INUNIT
INTEGER,                               INTENT(IN)  :: IOUT
[various types], DIMENSION (NR,NC),    INTENT(OUT) :: BUF
INTEGER,                               INTENT(OUT) :: IPRN
```

Explanation of arguments:

ANAME is text that describes the data to be read into the BUF array.

NR is the number of rows of data to be read and the first dimension of the BUF array.

NC is number of columns of data to be read and the second dimension of the BUF array.

INUNIT is the unit number from which data are to be read.

IOUT is the unit number associated with the file to which output is to be written, if required.

BUF is the array to be populated with data read from unit INUNIT. BUF may be of type DOUBLE PRECISION, REAL, or INTEGER.

IPRN is a flag that indicates whether to write the array and, if so, defines the format. IPRN is set to 0 when the value read from unit INUNIT exceeds the largest supported value. If IPRN is less than zero, the array is not written. If IPRN is greater than or equal to zero, UTL_READ2D calls UTL_WRITE2D to write the array. The supported formats and associated values of IPRN are shown in Table A- 4.

Subroutine UTL_READBLOCK

Description: This subroutine reads an input block of any of the supported formats (KEYWORDS, TABLE, or FILES). The data in the input block are stored in an input-block data structure accessible through a structure of type LLIST. The "Example Application: GROUP_EXAMPLE" section of Chapter 18 illustrates the use of UTL_READBLOCK.

Support for keyitems (see "Two forms of input blocks" in Chapter 3) and associated attributes that apply to each occurrence of the keyitem is defined by the value of the KITEM argument in the call to UTL_READBLOCK. The KITEM argument determines if a "keyitem" is defined for the input block. No keyitem is defined when KITEM has as its value a single asterisk ("*"); when KITEM has any other value, its value defines the keyitem.

Argument list:
```
(NDEFCOLS, BLABEL, COLNAMES, INUNIT, IOUT, KITEM, REQUIRED, HEAD, TAIL, KLISTS)
```

Declarations for argument-list variables:
```
INTEGER,                                 INTENT(IN)    :: NDEFCOLS
CHARACTER(LEN=*),                        INTENT(IN)    :: BLABEL
CHARACTER(LEN=20), DIMENSION(NDEFCOLSAR), INTENT(IN)   :: COLNAMES
INTEGER,                                 INTENT(IN)    :: INUNIT
INTEGER,                                 INTENT(IN)    :: IOUT
CHARACTER(LEN=*),                        INTENT(IN)    :: KITEM
LOGICAL,                                 INTENT(IN)    :: REQUIRED
TYPE (LLIST),                            POINTER       :: HEAD
TYPE (LLIST),                            POINTER       :: TAIL
INTEGER,                                 INTENT(INOUT) :: KLISTS
```

Explanation of arguments:
>   NDEFCOLS is the default number of columns, which applies when the TABLE format is specified and the COLUMNLABELS option is not specified.
>   BLABEL is a string of up to 20 characters, which specifies the blocklabel. The blocklabel is required in input to identify the block.
>   COLNAMES is an array of NDEFCOLS column names. The order in which the names are stored determines the default order for columns when the TABLE format is specified and the COLUMNLABELS option is not specified.
>   INUNIT is the unit number associated with the file from which the input block is to be read.
>   IOUT is the unit number associated with the file to which output is to be written.
>   KITEM is a string of up to 20 characters, which specifies the keyitem for the block. KITEM can be specified as the asterisk character ("*") to indicate that the block has no keyitem.
>   REQUIRED should be specified as .TRUE. if the block is to be required and .FALSE. otherwise.
>   HEAD is a pointer to type LLIST, which, if not already associated, will be assigned to point to the first LLIST type structure in the input-block data structure.
>   TAIL is a pointer to type LLIST, which will be assigned to point to the last LLIST type structure in the input-block data structure.
>   KLISTS is a counter, which is incremented by 1 for each list added to the nest of lists.

Notes:
1. See the "Input Blocks" section of Chapter 3 and the "Input-Block Data Types" section of Chapter 4 for details regarding reading input using this subroutine.
2. If REQUIRED is true and an input block identified with a blocklabel other than BLABEL is found on unit INUNIT, that input block is skipped over and the search for an input block for which the blocklabel is BLABEL continues. If REQUIRED is false and an input block identified with a blocklabel other than BLABEL is found on unit INUNIT, reading stops, and control returns to the calling program unit.
3. An input-block data structure containing a single linked list is generated when the KITEM argument is specified as an asterisk ("*").

4. An input-block data structure containing multiple linked lists is generated when the KITEM argument is not specified as an asterisk and the input block read by UTL_READBLOCK lists multiple occurrences of the keyitem.

Interface UTL_READMATRIX

Description: This generic subroutine interface will read matrix values in either the Complete Matrix or the Compressed Matrix format (see "Matrix Input and Output" section of Appendix A) and use the values to populate a structure of type CDMATRIX. This interface expects to read a line containing either "COMPLETEMATRIX" for the Complete Matrix format or "COMPRESSEDMATRIX" for the Compressed Matrix format and an optional array name of up to 12 characters.

Argument list:
```
(INUNIT, IOUT, RMAT)
```

Declarations for argument-list variables:
```
INTEGER,          INTENT(IN)    :: INUNIT
INTEGER,          INTENT(IN)    :: IOUT
TYPE (CDMATRIX), INTENT(INOUT) :: RMAT
```

Explanation of arguments:
> INUNIT is the unit number associated with the file from which the matrix is to be read.
> IOUT is the unit number associated with the file to which output is to be written.
> RMAT is the compressed matrix structure to be populated.

Notes:
1. Any information in RMAT when UTL_READMATRIX is called is lost. The array components of RMAT are deallocated and nullified before being reallocated and populated.
2. Comment lines (identified by a "#" symbol in column 1) and blank lines may precede the line containing "COMPLETEMATRIX" or "COMPRESSEDMATRIX" and are ignored.

Data Output Subroutines, Functions, and Generic Interfaces

This section documents subroutines, functions, and generic interfaces used primarily to write output or prepare data for output. The subroutines, functions, and interfaces are arranged alphabetically as follows.

> Subroutine UTL_COLLBL
> Subroutine UTL_COLNO
> Subroutine UTL_DX_WRITE_MCMV
> Subroutine UTL_DX_WRITE_PA
> Subroutine UTL_DX_WRITE_PCC
> Subroutine UTL_DX_WRITE_WT
> Subroutine UTL_EIGEN
> Subroutine UTL_PARSQMATRIX
> Function UTL_TABLESECTIONS
> Interface UTL_WRITE2D
> Subroutine UTL_WRITEBLOCK
> Subroutine UTL_WRITECDMATRIX
> Interface UTL_WRITEMATRIX

Subroutine UTL_WRITE_MESSAGE
Subroutine UTL_WRTSIG

Subroutine UTL_COLLBL

Description: This subroutine writes column names, as for a table or matrix heading.
Argument list:
```
(NCOL, NSPACE, NCPL, NDIG, IOUT, COLNAM)
```

Declarations for argument-list variables:
```
INTEGER,                               INTENT(IN) :: NCOL
INTEGER,                               INTENT(IN) :: NSPACE
INTEGER,                               INTENT(IN) :: NCPL
INTEGER,                               INTENT(IN) :: NDIG
INTEGER,                               INTENT(IN) :: IOUT
CHARACTER(LEN=*), DIMENSION(NCOL), INTENT(IN) :: COLNAM
```

Explanation of arguments:
NCOL is the number of column names and the dimension of the COLNAM array.
NSPACE is the number of blank spaces to leave at the start of the line.
NCPL is the number of column names to be written per line.
NDIG is the number of characters in each column field, exclusive of a space that will be used to separate columns.
IOUT is the unit number associated with the file to which output is to be written.
COLNAM contains the column names.

Subroutine UTL_COLNO

Description: This subroutine writes column numbers and a line of dots, as for a table or matrix heading.
Argument list:
```
(NLBL1, NLBL2, NSPACE, NCPL, NDIG, IOUT, NDOTS)
```

Declarations for argument-list variables:
```
INTEGER, INTENT(IN) :: NLBL1
INTEGER, INTENT(IN) :: NLBL2
INTEGER, INTENT(IN) :: NSPACE
INTEGER, INTENT(IN) :: NCPL
INTEGER, INTENT(IN) :: NDIG
INTEGER, INTENT(IN) :: IOUT
INTEGER, INTENT(IN), OPTIONAL :: NDOTS
```

Explanation of arguments:
NLBL1 is the start column number.
NLBL2 is the stop column number.
NSPACE is the number of blank spaces to leave at the start of the line.
NCPL is the number of column numbers per line.
NDIG is the number of characters in each column field, exclusive of a space that will be used to separate columns.
IOUT is the unit number associated with the file to which output is to be written.

NDOTS is the number of dots to add to the line of dots, below and after the last column number.

Subroutine UTL_DX_WRITE_MCMV

Description: This subroutine writes either an "_mc" file or an "_mv" data-exchange file (Chapter 17). The EXT argument determines which type of file is to be created. The name of the output file to be created is formed as OUTNAM concatenated with a period and the extension EXT.
Argument list:

```
(EXT, NPE, NPS, IPTR, OUTNAM, PARNAM, CMAT)
```

Declarations for argument-list variables:

```
CHARACTER(LEN=*),                          INTENT(IN) :: EXT
INTEGER,                                   INTENT(IN) :: NPE
INTEGER,                                   INTENT(IN) :: NPS
INTEGER,           DIMENSION(NPE),         INTENT(IN) :: IPTR
CHARACTER(LEN=*),                          INTENT(IN) :: OUTNAM
CHARACTER(LEN=12),DIMENSION(NPS),          INTENT(IN) :: PARNAM
DOUBLE PRECISION, DIMENSION(NPE,NPE),      INTENT(IN) :: CMAT
```

Explanation of arguments:

EXT is the extension to be used when opening the file, and it identifies the file type. Valid values for EXT are "_mc" and "_mv".

NPE is the number of adjustable parameters.

NPS is the total number of parameters.

IPTR contains, for each adjustable parameter, the position (element number) in the PARNAM array corresponding to the adjustable parameter.

OUTNAM is the root name for the data exchange files.

PARNAM contains the parameter names.

CMAT contains the matrix to be written to the output file. If EXT is "_mc", CMAT should contain the parameter correlation matrix. If EXT is "_mv", CMAT should contain the parameter covariance matrix.

Subroutine UTL_DX_WRITE_PA

Description: This subroutine writes data-exchange files of type _pa and _pasub (Table 17-5).
Argument list:

```
(ITERP, MAXITER, NPS, OUTNAM, PARNAM, PAREST, PVAL)
```

Declarations for argument-list variables:

```
INTEGER,                                         INTENT(IN) :: ITERP
INTEGER,                                         INTENT(IN) :: MAXITER
INTEGER,                                         INTENT(IN) :: NPS
CHARACTER(LEN=*),                                INTENT(IN) :: OUTNAM
CHARACTER(LEN=12), DIMENSION(NPS),               INTENT(IN) :: PARNAM
DOUBLE PRECISION, DIMENSION(MAXITER,NPS),        INTENT(IN) :: PAREST
DOUBLE PRECISION, DIMENSION(NPS), OPTIONAL,      INTENT(IN) :: PVAL
```

Explanation of arguments:

ITERP is parameter-estimation iteration number.

MAXITER is the maximum number of parameter-estimation iterations.

NPS is the number of parameters.
OUTNAM is the root name for the data exchange files.
PARNAM contains the parameter names.
PAREST contains the parameter estimates for each iteration.
PVAL contains the parameter estimates for a final parameter-estimation iteration, which, if present, are written to the _pa file.

Subroutine UTL_DX_WRITE_PCC

Description: This subroutine writes tables of large (≥ 0.85) parameter correlation coefficients to unit IOUT. If DATAEXCHANGE is true, it also writes the large parameter correlation coefficients to a _pcc data-exchange file (Chapter 17).
Argument list:
```
(NPE, NPS, BUFF, DATAEXCHANGE, IOUT, IPTR, OUTNAM, PARNAM)
```

Declarations for argument-list variables:
```
INTEGER,                                     INTENT(IN) :: NPE
INTEGER,                                     INTENT(IN) :: NPS
DOUBLE PRECISION, DIMENSION(NPE,NPE), INTENT(IN) :: BUFF
LOGICAL,                                     INTENT(IN) :: DATAEXCHANGE
INTEGER,                                     INTENT(IN) :: IOUT
INTEGER,           DIMENSION(NPE),    INTENT(IN) :: IPTR
CHARACTER(LEN=MAX_STRING_LEN),        INTENT(IN) :: OUTNAM
CHARACTER(LEN=12), DIMENSION(NPS),    INTENT(IN) :: PARNAM
```

Explanation of arguments:
NPE is the number of adjustable parameters.
NPS is the total number of parameters.
BUFF is an array (NPE,NPE) containing parameter correlation coefficients.
DATAEXCHANGE is true if data exchange files are to be written.
IOUT is the main output file unit number where the styled text will be written.
IPTR contains, for each adjustable parameter, the position (element number) in the PARNAM, PAREST, and PVAL arrays corresponding to the adjustable parameter.
OUTNAM is the root name for the data exchange files.
PARNAM contains the parameter names.

Subroutine UTL_DX_WRITE_WT

Description: This subroutine writes a data-exchange _wt or _wtpri file, which contains the full weight matrix for dependents or prior information in compressed matrix format, followed by its square root if the square root is in the argument list. The matrix data are written in COMPRESSEDMATRIX form (Appendix A).
Argument list:
```
(OUTNAM, EXT, WTFULL, WTFULLSQR)
```

Declarations for argument-list variables:
```
CHARACTER(LEN=*),            INTENT(IN) :: OUTNAM
CHARACTER(LEN=*),            INTENT(IN) :: EXT
TYPE(CDMATRIX),              INTENT(IN) :: WTFULL
TYPE(CDMATRIX), OPTIONAL, INTENT(IN) :: WTFULLSQR
```

Explanation of arguments:

OUTNAM is the root file name for the output file.

EXT is one of two supported data-exchange file types to be created; its value may be either "_wt" or "_wtpri." If EXT is "_wt", matrix(ces) for dependents is (are) written to a data-exchange file of type _wt. If EXT is "_wtpri," matrix(ces) for prior information is (are) written to a data-exchange file of type _wtpri (Table 17-3).

WTFULL contains the full weight matrix for dependents or prior information in compressed matrix format. If EXT is "_wt," WTFULL is assumed to contain a full weight matrix for dependents. If EXT is "_wtpri," WTFULL is assumed to contain a full weight matrix for prior information.

WTFULLSQR, if present, contains the square root of the full weight matrix for dependents or prior information in compressed matrix format. If EXT is "_wt," WTFULLSQR is assumed to contain the square root of the full weight matrix for dependents. If EXT is "_wtpri," WTFULLSQR is assumed to contain the square root of the full weight matrix for prior information.

Subroutine UTL_EIGEN

Description: Calculate eigenvalues and eigenvectors of a square matrix. Write an eigenvalue for each of NPE parameters and an NPExNPE table of eigenvectors to output.

Argument list:

```
(IOUT, IPRC, NPE, NPS, IPTR, ITERP, PARNAM, PVALINIT, PVAL, C)
```

Declarations for argument-list variables:

```
INTEGER,                                    INTENT(IN)    :: IOUT
INTEGER,                                    INTENT(IN)    :: IPRC
INTEGER,                                    INTENT(IN)    :: NPE
INTEGER,                                    INTENT(IN)    :: NPS
INTEGER,            DIMENSION(NPE),         INTENT(IN)    :: IPTR
INTEGER,                                    INTENT(IN)    :: ITERP
CHARACTER(LEN=12),  DIMENSION(NPS),         INTENT(IN)    :: PARNAM
DOUBLE PRECISION,   DIMENSION(NPS),         INTENT(IN)    :: PVALINIT
DOUBLE PRECISION,   DIMENSION(NPS),         INTENT(IN)    :: PVAL
DOUBLE PRECISION,   DIMENSION(NPE,NPE),     INTENT(INOUT) :: C
```

Explanation of arguments:

IOUT is the unit number of the output file where the eigenvalues and eigenvectors will be written.

IPRC is a code that defines the format to be used when is written. Valid values of IPRC and corresponding output formats are documented in the description of UTL_WRITEMATRIX, below.

NPE is the number of adjustable parameters.

NPS is the total number of parameters, including adjustable and non-adjustable, primary or derived parameters.

IPTR contains, for each adjustable parameter, the position (element number) in the PARNAM, PVALINIT, and PVAL arrays corresponding to the adjustable parameter.

ITERP is the current iteration number.

PARNAM contains the parameter names.

PVALINIT contains the initial parameter values.

PVAL contains the current parameter values.

C is the variance/covariance matrix of the parameters.

Note:

In this routine, C is scaled with the parameters, then the eigenvectors and eigenvalues are calculated.

Subroutine UTL_PARSQMATRIX

Description: Writes a square variance-covariance matrix labeled with parameter names.

Argument list:

```
(IOUT, IPRC, NPE, NPS, BUF, IPTR, PARNAM)
```

Declarations for argument-list variables:

```
INTEGER,                                 INTENT(IN) :: IOUT
INTEGER,                                 INTENT(IN) :: IPRC
INTEGER,                                 INTENT(IN) :: NPE
INTEGER,                                 INTENT(IN) :: NPS
DOUBLE PRECISION,   DIMENSION(NPE,NPE),  INTENT(IN) :: BUF
INTEGER,            DIMENSION(NPE),      INTENT(IN) :: IPTR
CHARACTER(LEN=12),  DIMENSION(NPS),      INTENT(IN) :: PARNAM
```

Explanation of arguments:

IOUT is the unit number associated with the file to which output is to be written.

IPRC is a code that defines the format to be used when BUF is written. Valid values of IPRC and corresponding output formats are documented in the description of subroutine UTL_WRITEMATRIX.

NPE is the number of adjustable parameters.

NPS is the total number of parameters, including adjustable and non-adjustable, primary or derived parameters.

BUF is the variance-covariance matrix of the parameters.

IPTR contains, for each adjustable parameter, the position (element number) in the PARNAM, PVALINIT, and PVAL arrays corresponding to the adjustable parameter.

PARNAM contains the parameter names.

Function UTL_TABLESECTIONS

Description: This function returns the number of table sections required to write a table.

Argument list:

```
(NCOLUMNS, NCSECTION)
```

Result:

```
NSECTS
```

Declarations for argument-list and result variables:

```
INTEGER, INTENT(IN) :: NCOLUMNS
INTEGER, INTENT(IN) :: NCSECTION
INTEGER             :: NSECTS
```

Explanation of arguments:

NCOLUMNS is the number of columns in entire table.

154

NCSECTION is the number of columns per table section.

NSECTS is the number of table sections.

Interface UTL_WRITE2D

Description: This generic subroutine interface writes a 2-D array to unit IOUT.

Argument list:

```
(IOUT, IPRN, NR, NC, BUF, TEXT)
```

Declarations for argument-list variables:

```
INTEGER,                              INTENT(IN) :: IOUT
INTEGER,                              INTENT(IN) :: IPRN
INTEGER,                              INTENT(IN) :: NR
INTEGER,                              INTENT(IN) :: NC
[various types], DIMENSION(NR,NC),    INTENT(IN) :: BUF
CHARACTER(LEN=*),                     INTENT(IN) :: TEXT
```

Explanation of arguments:

IOUT is the unit number associated with the file to which output is to be written.

IPRN is a code that defines the format to be used when BUF is written. Valid values of IPRN and corresponding output formats are shown in Table A-4.

NR is the first dimension of BUF.

NC is the second dimension of BUF.

BUF contains the data to be written to unit IOUT. BUF may be of type DOUBLE PRECISION, REAL, or INTEGER.

TEXT is text that describes the contents of BUF.

Subroutine UTL_WRITEBLOCK

Description: This subroutine writes a table of keywords and values for all entries in an input-block data structure.

Argument list:

```
(HEAD, IOUT)
```

Declarations for argument-list variables:

```
TYPE (LLIST), POINTER    :: HEAD
INTEGER,      INTENT(IN) :: IOUT
```

Explanation of arguments:

HEAD points to the first LLIST structure of an input-block data structure.

IOUT is the unit number associated with the file to which output is to be written.

Note:

This subroutine is intended mainly for debugging purposes.

Subroutine UTL_WRITECDMATRIX

Description: This subroutine writes the contents of a CDMATRIX structure. If the ARRAYNAME component of the structure is not blank, the value of ARRAYNAME is included in the first line of output.

Appendix D: Utilities (UTL) Module of Chapter 6

Argument list:
```
(CDM, IFLAG, IOUT, INPUTFORMAT, COLNAM, ROWNAM)
```

Declarations for argument-list variables:
```
TYPE (CDMATRIX),                                INTENT(IN) :: CDM
INTEGER,                                         INTENT(IN) :: IFLAG
INTEGER,                                         INTENT(IN) :: IOUT
LOGICAL,                             OPTIONAL, INTENT(IN) :: INPUTFORMAT
CHARACTER(LEN=*), DIMENSION(CDM%NC), OPTIONAL, INTENT(IN) :: COLNAM
CHARACTER(LEN=*), DIMENSION(CDM%NR), OPTIONAL, INTENT(IN) :: ROWNAM
```

Explanation of arguments:
> CDM is the CDMATRIX structure to be written.
> IFLAG is a flag that indicates, in conjunction with INPUTFORMAT, the format in which the contents of CDM are to be written. See the notes below for details.
> IOUT is the unit number associated with the file to which output is to be written.
> INPUTFORMAT, in conjunction with IFLAG, indicates the format to be used in writing the matrix. If INPUTFORMAT is absent, it is by default assumed to be false. See the notes below for details.
> COLNAM contains the column names.
> ROWNAM contains the row names.

Notes:
1. If INPUTFORMAT is true and IFLAG is less than 1, the matrix is written in the "Compressed matrix" format described in Appendix A, starting with the string "COMPRESSEDMATRIX". This format is suitable for input intended to be read by UTL_READMATRIX.
2. If INPUTFORMAT is true and IFLAG is greater than or equal to 1, the matrix is written in the "Complete matrix" format described in Appendix A, starting with the string "COMPLETEMATRIX". This format is suitable for input intended to be read by UTL_READMATRIX.
3. If INPUTFORMAT is false and IFLAG is less than 1, the matrix is written in its compressed form, which is intended primarily for debugging purposes. In this form, the values of CDM components IDIM, NNZ, NR, and NC are written, followed by tables showing the contents of the IPOS, DVAL, and ICOL arrays.
4. If INPUTFORMAT is false and IFLAG is greater than or equal to 1, the full matrix is written in row and column form. If COLNAM and ROWNAM are present, the columns and rows are labeled with the names contained in those arrays. If COLNAM and ROWNAM are absent, the columns and row are labeled with numbers. If COLNAM is present and ROWNAM is absent, the columns are labeled with the names in COLNAM and the rows are labeled with numbers.

Interface UTL_WRITEMATRIX

Description: This generic subroutine interface writes a 2-D matrix with row and column names.
Argument list:
```
(NROW, NCOL, BUF, ROWNAM, COLNAM, IPRC, IOUT)
```

Declarations for argument-list variables:
```
INTEGER,                                INTENT(IN) :: NROW
INTEGER,                                INTENT(IN) :: NCOL
```

156

```
[various types], DIMENSION(NROW,NCOL),   INTENT(IN) :: BUF
CHARACTER(LEN=*), DIMENSION(NROW),       INTENT(IN) :: ROWNAM
CHARACTER(LEN=*), DIMENSION(NCOL),       INTENT(IN) :: COLNAM
INTEGER,                                 INTENT(IN) :: IPRC
INTEGER,                                 INTENT(IN) :: IOUT
```

Explanation of arguments:

NROW is the first dimension of BUF.

NCOL is the second dimension of BUF.

BUF is the 2-D array to be written; it may be of type DOUBLE PRECISION or REAL.

ROWNAM contains the row names.

COLNAM contains the column names.

IPRC is a code that defines the format to be used when BUF is written. Valid values of IPRC and corresponding output formats are described in the note, below.

IOUT is the unit number associated with the file to which output is to be written.

Note:

The format associated with IPRC=0 is used when IPRC is less than zero or exceeds 10. The formats associated with the supported values of IPRC are as follows:

IPRC	Format
0	11G10.3
1	11G10.3
2	10G11.4
3	9G12.5
4	8G13.6
5	8G14.7
6	6G10.3
7	5G11.4
8	5G12.5
9	4G13.6
10	4G14.7

Subroutine UTL_WRITE_MESSAGE

Description: This subroutine formats and writes the message contained in AMESSAGE of the Global Data Module. If AMESSAGE is blank, UTL_WRITE_MESSAGE returns without writing anything. If the message is longer than 80 characters, it is written on multiple lines, with line breaks between space-delimited words.

Argument list:
```
(IUNIT, ERROR, LEADSPACE, ENDSPACE)
```

Declarations for argument-list variables:
```
INTEGER, INTENT(IN), OPTIONAL            ::IUNIT
CHARACTER (LEN=*), INTENT(IN), OPTIONAL ::ERROR
CHARACTER (LEN=*), INTENT(IN), OPTIONAL ::LEADSPACE
CHARACTER (LEN=*), INTENT(IN), OPTIONAL ::ENDSPACE
```

Explanation of arguments:

IUNIT, if present and if greater than or equal to 1, is the unit number associated with an open file to which output is to be written. If IUNIT is absent or if IUNIT is less than 1, output is written to the screen.

ERROR, if present and equal to "yes" (in any combination of uppercase or lowercase letters), indicates that the message should be prefixed by "Error: ".

LEADSPACE, if present and equal to "yes" (in any combination of uppercase or lowercase letters), indicates that a blank line should be printed before the message.

ENDSPACE, if present and equal to "yes" (in any combination of uppercase or lowercase letters), indicates that a blank line should be printed after the message.

Subroutine UTL_WRTSIG

Description: This subroutine writes a number to a character variable within a specified field width with maximum precision.

Argument list:
```
(IFAIL, VAL, WORD, NW, PRECPR, TVAL, NOPNT)
```

Declarations for argument-list variables:
```
INTEGER,           INTENT(OUT) :: IFAIL
DOUBLE PRECISION,  INTENT(IN)  :: VAL
CHARACTER(LEN=*),  INTENT(OUT) :: WORD
INTEGER,           INTENT(IN)  :: NW
INTEGER,           INTENT(IN)  :: PRECPR
DOUBLE PRECISION,  INTENT(OUT) :: TVAL
INTEGER,           INTENT(IN)  :: NOPNT
```

Explanation of arguments:

IFAIL is failure criterion:

IFAIL = 0 – Successful completion of subroutine

IFAIL = 1 – Number too large or small for single precision type

IFAIL = 2 – Number too large or small for double precision type

IFAIL = 3 – Field width too small to represent number

IFAIL < 0 – Error internal to UTL_WRTSIG

VAL is the value to be written.

WORD is the character representation of precision-limited value.

NW is the available field width.

PRECPR is a flag indicating the precision protocol:

PRECPR = 0 – WORD will contain single-precision representation of TVAL, with exponent in "E" format when required.

PRECPR = 1 – WORD will contain double-precision representation of TVAL, with exponent in "E" format when required.

PRECPR = 2 – WORD will contain double-precision representation of TVAL, with exponent in "D" format when required.

TVAL is the value as written, truncated to available precision.

NOPNT is a flag indicating the decimal-point protocol:

NOPNT = 0 – Decimal point is required.

NOPNT = 1 – Decimal point is optional, and will be omitted if doing so allows an additional significant figure to be written.

Appendix D: Utilities (UTL) Module of Chapter 6

Data Manipulation Subroutines – Statistical

This section documents four subroutines used primarily to calculate statistics. The subroutines are arranged alphabetically as follows.

Subroutine UTL_CALCWT
Subroutine UTL_CHISQ
Subroutine UTL_FSTT
Subroutine UTL_STUD_T

Subroutine UTL_CALCWT

Description: This subroutine calculates the weight and variance associated with an observed or reference value, using a statistic and a flag indicating the type of statistic from which the weight and variance are to be calculated.

Argument list:

```
(IOUT, NAME, STATFLAG, STATISTIC, VALUE, WTMULT, IERR, WEIGHT, VAR)
```

Declarations for argument-list variables:

```
INTEGER,                 INTENT(IN)   :: IOUT
CHARACTER(LEN=LENDNAM),  INTENT(IN)   :: NAME
CHARACTER(LEN=6),        INTENT(IN)   :: STATFLAG
DOUBLE PRECISION,        INTENT(IN)   :: STATISTIC
DOUBLE PRECISION,        INTENT(IN)   :: VALUE
DOUBLE PRECISION,        INTENT(IN)   :: WTMULT
INTEGER,                 INTENT(INOUT) :: IERR
DOUBLE PRECISION,        INTENT(OUT)  :: WEIGHT
DOUBLE PRECISION,        INTENT(OUT)  :: VAR
```

Explanation of arguments:

IOUT is the unit number associated with the file to which error messages are to be written, if required.

NAME is the name associated with the observed or reference value.

STATFLAG indicates the type of statistic provided in the STATISTIC argument. The options are:

"CV" – Coefficient of variation

"SD" – Standard deviation

"VAR" – Variance

"WT" – Weight

"SQRWT" – Square root of the weight

STATISTIC is the value from which the weight is to be calculated.

VALUE is the observed or reference value to which the weight is to apply.

WTMULT is a value that will be used to multiply the weight.

IERR is incremented by 1 if an error is encountered in the subroutine.

WEIGHT is the weight.

VAR is the variance.

Notes:

1. The formula used to calculate the weight is determined by the value of STATFLAG, as listed in Table 10-1.

2. Variance is calculated as the inverse of the weight.

Appendix D: Utilities (UTL) Module of Chapter 6

Subroutine UTL_CHISQ

Description: Determine Chi-squared statistic for a specified number of degrees of freedom for a two-sided significance level of 0.05.
Argument list:
```
(IDOF,CHISQL,CHISQU)
```

Declarations for argument-list variables:
```
INTEGER,           INTENT(IN)  :: IDOF
DOUBLE PRECISION, INTENT(OUT) :: CHISQL
DOUBLE PRECISION, INTENT(OUT) :: CHISQU
```

Explanation of arguments:
   IDOF is the number of degrees of freedom.
   CHISQL is value required to calculate the lower confidence interval=(DOF * VAR / CHISQL).
   CHISQU is value required to calculate the upper confidence interval=(DOF * VAR / CHISQU).

Subroutine UTL_FSTT

Description: Determine the value of the F statistic for 5 percentage points.
Argument list:
```
(NP, IDOF, TST)
```

Declarations for argument-list variables:
```
INTEGER,           INTENT(IN)  :: NP
INTEGER,           INTENT(IN)  :: IDOF
DOUBLE PRECISION, INTENT(OUT) :: TST
```

Explanation of arguments:
   NP is the number of parameters.
   IDOF is the degrees of freedom.
   TST is the value of the F statistic.

Subroutine UTL_STUD_T

Description: This subroutine determines the value of the Student's t statistic needed to calculate linear individual confidence intervals for a two-sided significance level.
Argument list:
```
(IDOF, TST)
```

Declarations for argument-list variables:
```
INTEGER,           INTENT(IN)  :: IDOF
DOUBLE PRECISION, INTENT(OUT) :: TST
```

Explanation of arguments:
   IDOF is the number of degrees of freedom (NOBS – NPE – MPR).
   TST is the t-statistic.

Data Manipulation Subroutines, Functions, and Generic Interfaces – Character and String Management

This section documents subroutines, functions, and generic interfaces used primarily for character and string management. The subroutines, functions, and interfaces are arranged alphabetically as follows.

Subroutine UTL_ADDQUOTE
Subroutine UTL_ARR2STRING
Subroutine UTL_CASE
Subroutine UTL_CASETRANS
Interface UTL_CHAR2NUM
Subroutine UTL_CHECK_NAMES
Subroutine UTL_COMPRESSLINE
Function UTL_COUNTSUBS
Subroutine UTL_GETTOKEN
Subroutine UTL_GETWORD
Subroutine UTL_GETWORDLEFT
Interface UTL_NUM2CHAR
Subroutine UTL_REMCHAR
Interface UTL_RWORD
Function UTL_SAMENAME
Subroutine UTL_SHELLSORT
Subroutine UTL_STRING2ARR
Subroutine UTL_TABREP
Subroutine UTL_WHICH1

Subroutine UTL_ADDQUOTE

Description: This subroutine adds quotes to a text string if it has one or more spaces embedded in it.
Argument list:
```
(TEXTIN, TEXTOUT)
```

Declarations for argument-list variables:
```
CHARACTER (LEN=*), INTENT(IN)   :: TEXTIN
CHARACTER (LEN=*), INTENT(OUT)  :: TEXTOUT
```

Explanation of arguments:
    TEXTIN is the input text string.
    TEXTOUT is the output text string, with double quotes added if TEXTIN includes one or more embedded spaces.

Subroutine UTL_ARR2STRING

Description: This subroutine assigns the contents of a character (LEN=1) array to a string.
Argument list:
```
(IDIM, CHARARR, STRING)
```

Appendix D: Utilities (UTL) Module of Chapter 6

Declarations for argument-list variables:
```
INTEGER,                                INTENT(IN)  :: IDIM
CHARACTER(LEN=1), DIMENSION(IDIM), INTENT(IN)  :: CHARARR
CHARACTER(LEN=*),                       INTENT(OUT) :: STRING
```

Explanation of arguments:
   IDIM is the dimension of the CHARARR array.
   CHARARR is the array to be assigned to the string.
   STRING is the character string to be populated.
Note:
   If IDIM exceeds the length of STRING, STRING is populated using the elements of
      CHARARR starting at the first element, and elements beyond the length of STRING are
      truncated.

Subroutine UTL_CASE

Description: This subroutine copies a character string and performs a case conversion on the copy.
Argument list:
```
(WORDIN, WORDOUT, ICASE)
```

Declarations for argument-list variables:
```
CHARACTER(LEN=*),   INTENT(IN)  :: WORDIN
CHARACTER(LEN=*),   INTENT(OUT) :: WORDOUT
INTEGER, OPTIONAL, INTENT(IN)  :: ICASE
```

Explanation of arguments:
   WORDIN is the input string.
   WORDOUT is the output string.
   ICASE is a flag that indicates what kind of case conversion to perform. If ICASE is absent, it
      defaults to 1 and the string is converted to all uppercase. The values and meanings are:
      ICASE < 0 – Convert the string to all lowercase.
      ICASE = 0 – No case conversion is performed.
      ICASE > 0 – Convert the string to all uppercase.

Subroutine UTL_CASETRANS

Description: This subroutine converts a character string to all uppercase or all lowercase.
Argument list:
```
(STRING, HI_OR_LO)
```

Declarations for argument-list variables:
```
CHARACTER (LEN=*), INTENT(INOUT) :: STRING
CHARACTER (LEN=*), INTENT(IN)     :: HI_OR_LO
```

Explanation of arguments:
   STRING is the character string to be converted.
   HI_OR_LO is a flag indicating whether the conversion is to be to all uppercase or to all
      lowercase. The values and meanings are:
      HI_OR_LO = "lo" – Convert STRING to lowercase.
      HI_OR_LO = "hi" – Convert STRING to uppercase.

Appendix D: Utilities (UTL) Module of Chapter 6

Interface UTL_CHAR2NUM

Description: This generic subroutine interface converts a character string to either an INTEGER, a
REAL number, a DOUBLE PRECISION number, or a LOGICAL variable.
Argument list:
```
(IFAIL, STRING, NUM)
```

Declarations for argument-list variables:
```
INTEGER,            INTENT(OUT) :: IFAIL
CHARACTER (LEN=*),  INTENT(IN)  :: STRING
[various types],    INTENT(OUT) :: NUM
```

Explanation of arguments:
> IFAIL is an error flag. Any value other that 0 indicates an error has occurred.
> STRING is the character string to be converted to a number.
> NUM is the numeric value read from STRING. NUM may be of type INTEGER, REAL,
> DOUBLE PRECISION, or LOGICAL.

Note:
> If NUM is of type LOGICAL, any string starting with "T" or ".T" is interpreted as true, and any
> string starting with "F" or ".F" is interpreted as false; the T or F can be either uppercase or
> lowercase.

Subroutine UTL_CHECK_NAMES

Description: This subroutine checks an array of names for conformance with the JUPITER naming
convention. It also left-justifies the name in each element of the array.
Argument list:
```
(IOUT, NUMNAMES, NAMEARRAY)
```

Declarations for argument-list variables:
```
INTEGER,                                    INTENT(IN)    :: IOUT
INTEGER,                                    INTENT(IN)    :: NUMNAMES
CHARACTER(LEN=*), DIMENSION(NUMNAMES),      INTENT(INOUT) :: NAMEARRAY
```

Explanation of arguments:
> IOUT is the unit number associated with the file to which output is to be written.
> NUMNAMES is the declared dimension of the NAMEARRAY array.
> NAMEARRAY contains the names to be checked.

Subroutine UTL_COMPRESSLINE

Description: Compress an instruction line by replacing multiple, consecutive blank characters with
single blank characters.
Argument list:
```
(LINE)
```

Declarations for argument-list variables:
```
CHARACTER (LEN=*), INTENT(INOUT) :: LINE
```

Explanation of arguments:

    LINE is a string of text to be compressed.

Function UTL_COUNTSUBS

Description: This function returns the number of occurrences of substring SUBSTR in string STR; the comparison is case-sensitive.

Argument list:
```
(STR, SUBSTR)
```

Result:
```
K
```

Declarations for argument-list and result variables:
```
CHARACTER(LEN=*), INTENT(IN) :: STR
CHARACTER(LEN=*), INTENT(IN) :: SUBSTR
INTEGER                      :: K
```

Explanation of arguments and result:

    STR is the string in which occurrences of the substring are counted.

    SUBSTR is the substring of interest.

    K is assigned the number of occurrences of SUBSTR in STR.

Notes:

    1. If the length of SUBSTR is greater than that of STR, the result value is 0.

    2. Overlapping occurrences of SUBSTR in STR contribute to the result. For example, if STR is "aaa" and SUBSTR is "aa", the result is 2.

Subroutine UTL_GETTOKEN

Description: This subroutine extracts a token (either a word or a phrase) from a character string. Words are separated by a comma, a tab character, or one or more spaces. Words containing unmatched single (') or double (") quotes will cause additional, following words to be included in the token until all quotes are matched. The search for the token begins at the starting pointer (ICOL). Unmatched single or double quotes may not be embedded in TOKEN.

Argument list:
```
(KERR, INUNIT, IOUT, NCODE, ICOL, LINE, ISTART, ISTOP, TOKEN)
```

Declarations for argument-list variables:
```
INTEGER,          INTENT(INOUT) :: KERR
INTEGER,          INTENT(IN)    :: INUNIT
INTEGER,          INTENT(IN)    :: IOUT
INTEGER,          INTENT(IN)    :: NCODE
INTEGER,          INTENT(INOUT) :: ICOL
CHARACTER(LEN=*), INTENT(INOUT) :: LINE
INTEGER,          INTENT(OUT)   :: ISTART
INTEGER,          INTENT(OUT)   :: ISTOP
CHARACTER(LEN=*), INTENT(OUT)   :: TOKEN
```

Explanation of arguments:

    KERR is incremented by one if all single and double quotes cannot be matched.

INUNIT, if provided as a value greater than zero, is the unit number from which LINE was read; if INUNIT is greater than zero and an error is encountered, an error message is written to unit IOUT identifying INUNIT as the location of the error. If INUNIT is 0 or less, the error message does not identify the location of the error.

IOUT is the unit number associated with the output file to which error messages are to be written.

NCODE is a flag that is used to determine if the returned TOKEN is to be converted to uppercase. If NCODE equals 1, TOKEN and the alphabetic characters in LINE between positions ISTART and ISTOP are converted to uppercase.

ICOL is the position in LINE from which the search for TOKEN is started. On return, ICOL is assigned the next position after ISTOP.

LINE is the string from which TOKEN is to be extracted.

ISTART is the starting position in LINE of TOKEN, including enclosing quotes, if any.

ISTOP is the ending position in LINE of TOKEN, including enclosing quotes, if any.

TOKEN is assigned as the substring in LINE starting at ICOL, which is delimited by spaces, commas, or tabs and defined by matching single and double quotes.

Notes:
1. ISTART and ISTOP are assigned the starting and ending positions in LINE containing TOKEN. If TOKEN is enclosed in single or double quotes in LINE, the string defined by ISTART and ISTOP includes the initial and final quotes.
2. If ISTART and ISTOP point to matching single or double quotes, TOKEN is returned without these initial and final quotes. Other (embedded) matching quotes are retained in TOKEN.

Subroutine UTL_GETWORD

Description: This subroutine defines the start and end of the next space-delimited word (after a specified position) in a string.

Argument list:
```
(J1, LINE, NBLC, IFAIL, NUM1, NUM2)
```

Declarations for argument-list variables:
```
INTEGER,              INTENT(IN)  :: J1
CHARACTER (LEN=*),    INTENT(IN)  :: LINE
INTEGER,              INTENT(IN)  :: NBLC
INTEGER,              INTENT(OUT) :: IFAIL
INTEGER,              INTENT(OUT) :: NUM1
INTEGER,              INTENT(OUT) :: NUM2
```

Explanation of arguments:
J1 is the position in LINE after which to search for a word.

LINE is the input character string.

NBLC is the last position in LINE in which to search for a word.

IFAIL is a flag that is set to 0 if no errors are encountered, or to the number of errors if any errors are encountered.

NUM1 is the position in LINE of the beginning of the located word.

NUM2 is the position in LINE of the end of the located word.

Appendix D: Utilities (UTL) Module of Chapter 6

Subroutine UTL_GETWORDLEFT

Description: This subroutine assigns to TPAR the left-justified contents of the substring of LINE between J1 and J2.

Argument list:
```
(LINE, IFAIL, J1, J2, TPAR)
```

Declarations for argument-list variables:
```
CHARACTER (LEN=*), INTENT(IN)  :: LINE
INTEGER, INTENT(OUT)           :: IFAIL
INTEGER, INTENT(IN)            :: J1
INTEGER, INTENT(IN)            :: J2
CHARACTER (LEN=*), INTENT(OUT) :: TPAR
```

Explanation of arguments:

LINE is the input character string.

IFAIL is a flag that is set to 0 if no errors are encountered, or to the number of errors if any errors are encountered.

J1 is the leftmost position in LINE to search for a word.

J2 is the rightmost position in LINE to search for a word.

TPAR is assigned the contents of LINE from the first non-blank character to the right of J1, up to the length of TPAR or up to J2.

Interface UTL_NUM2CHAR

Description: This generic subroutine interface converts a number to a character string.

Argument list:
```
(VALUE, STRING, NCHAR)
```

Declarations for argument-list variables:
```
[various types],    INTENT(IN)  :: VALUE
CHARACTER (LEN=*), INTENT(OUT)  :: STRING
INTEGER, OPTIONAL, INTENT(IN)   :: NCHAR
```

Explanation of arguments:

VALUE is the number to be converted to a character string. VALUE may be of type INTEGER, REAL, or DOUBLE PRECISION.

STRING is a character representation of VALUE.

NCHAR is the number of characters of STRING to be used in the character representation.

Subroutine UTL_REMCHAR

Description: This subroutine replaces all occurrences of one character string in another character string with blanks.

Argument list:
```
(ASTRING, ACH)
```

Declarations for argument-list variables:
```
CHARACTER(LEN=*), INTENT(INOUT) :: ASTRING
CHARACTER(LEN=*), INTENT(IN)     :: ACH
```

Explanation of arguments:

ASTRING is the character string to be modified.

ACH is the character string which, when found in ASTRING, is to be replaced by blanks.

Notes:

1. All occurrences of ACH in ASTRING are replaced with blanks in the same position(s) originally occupied by ACH.

2. After replacements are performed, the non-blank text in ASTRING is left-justified.

Interface UTL_RWORD

Description: This generic subroutine interface locates a "word" in a line of text and optionally converts the word to uppercase or to a number.

Argument list:

```
(IOUT, NCODE, RECQUOTES, ICOL, LINE, ISTART, ISTOP, N, R, INUNIT)
```

Declarations for argument-list variables:

```
INTEGER,            INTENT(IN)    :: IOUT
INTEGER,            INTENT(IN)    :: NCODE
LOGICAL,            INTENT(IN)    :: RECQUOTES
INTEGER,            INTENT(INOUT) :: ICOL
CHARACTER(LEN=*),   INTENT(INOUT) :: LINE
INTEGER,            INTENT(OUT)   :: ISTART
INTEGER,            INTENT(OUT)   :: ISTOP
INTEGER,            INTENT(OUT)   :: N
[various types],    INTENT(OUT)   :: R
INTEGER, OPTIONAL,  INTENT(IN)    :: INUNIT
```

Explanation of arguments:

IOUT is the unit number to which error messages, if any, are to be written. If IOUT is 0, output is written to default output; if IOUT is less than 1, no output is written.

NCODE is a flag that determines what conversion, if any, should be performed on the word. Values and the resulting conversions are:

NCODE=1 – Convert the word to uppercase.

NCODE=2 – Convert the word to an integer and return as argument N.

NCODE=3 – Convert the word to either a REAL or a DOUBLE PRECISION number, according to the type of argument R.

Any other value of NCODE results in no conversion.

RECQUOTES controls recognition of single (') and double (") quotes as delimiters.

If RECQUOTES is true, single and double quotes in LINE will be interpreted as delimiters. For a word that begins with a quote, the word starts with the character after the quote and ends with the character preceding a subsequent, matching quote. Thus, a quoted word can include spaces, commas, and tab characters. The quoted word cannot contain an embedded quote character that matches the initial quote character.

If RECQUOTES is false, single and double quotes are treated as ordinary characters, and spaces, tab characters, and commas are interpreted as delimiters.

ICOL is the position in LINE at which the search for a word begins. On return, ICOL is the position following ISTOP, or if RECQUOTES is true and the word is quoted, ICOL is the position following the final quote delimiting the word.

LINE is the text string from which a word is to be extracted.

ISTART is the position in LINE of the beginning of the word.

ISTOP is the position in LINE of the end of the word.

N is the numeric value of the word, interpreted as an INTEGER.

R is the numeric value of the word, interpreted as a floating-point number. R is returned as either a REAL or a DOUBLE PRECISION value, depending on the declared type of the variable in the CALL statement that corresponds to R.

INUNIT, if greater than 0, is the unit number corresponding to the file from which LINE was read.

Notes:

1. ISTART and ISTOP will be returned with the starting and ending character positions of the word.

2. The last character in LINE is set to blank so that if any problems occur with finding a word, ISTART and ISTOP will point to this blank character. Thus, a word will always be returned unless there is a numeric conversion error. Be sure that the last character in LINE is not an important character because it will always be set to blank.

3. A word starts with the first character that is not a space, comma, or tab character and ends when a subsequent character is a space, comma, or TAB character. Note that these parsing rules do not treat two commas separated by one or more spaces as a null word.

4. Number conversion error is written to unit IOUT if IOUT is positive; error is written to default output if IOUT is 0. No error message is written if IOUT is negative.

5. If INUNIT is present and greater than 0, error messages identify INUNIT as the unit number from which LINE was read.

Function UTL_SAMENAME

Description: This function performs a case-insensitive comparison of two strings.
Argument list:
```
(NAME1, NAME2)
```

Result:
```
SAME
```

Declarations for argument-list and result variables:
```
CHARACTER(LEN=*), INTENT(IN) :: NAME1
CHARACTER(LEN=*), INTENT(IN) :: NAME2
LOGICAL                      :: SAME
```

Explanation of arguments:

NAME1 and NAME2 are the strings to be compared.

SAME is the result of the case-insensitive comparison.

Note:

Copies of NAME1 and NAME2 are converted to uppercase and compared. If the converted strings are identical, SAME is returned as TRUE; otherwise, SAME is returned as FALSE.

Subroutine UTL_SHELLSORT

Description: Sort values stored in array X in ascending order. The sort order is determined by the numeric ASCII codes for characters. Note that in the ASCII order, all capital letters precede all lowercase letters. Blank characters in the strings are not ignored.

Argument list:
```
(N, X)
```

Declarations for argument-list variables:
```
INTEGER,                           INTENT(IN)    ::  N
CHARACTER(LEN=*), DIMENSION(N), INTENT(INOUT) ::  X
```

Explanation of arguments:
> N is the number of values in array X.
> X contains the values to be sorted.

Note:
> Algorithm obtained from National Institute of Standards and Technology, accessed on January 18, 2006 at <http://www.nist.gov/dads/HTML/shellsort.html>. The method was described by Donald L. Shell of General Electric.

Subroutine UTL_STRING2ARR

Description: This subroutine assigns the contents of a string to a character (LEN=1) array.

Argument list:
```
(IDIM, STRING, CHARARR)
```

Declarations for argument-list variables:
```
INTEGER,                           INTENT(IN)  :: IDIM
CHARACTER(LEN=*),                  INTENT(IN)  :: STRING
CHARACTER(LEN=1), DIMENSION(IDIM), INTENT(OUT) :: CHARARR
```

Explanation of arguments:
> IDIM is the dimension of the CHARARR array.
> STRING is the character string to be assigned to the array.
> CHARARR is the array to be populated.

Note:
> If the length of STRING exceeds IDIM, the first IDIM characters in STRING are stored in CHARARR.

Subroutine UTL_TABREP

Description: Replace <TAB> characters in a string with blanks such that characters following <TAB> characters start in columns 1, 9, 17, and so on.

Argument list:
```
(LINE)
```

Declaration for argument-list variable:
```
CHARACTER (LEN=*), INTENT(INOUT) :: LINE
```

Explanation of arguments:
> LINE is the character string to have <TAB> characters replaced with blanks.

Subroutine UTL_WHICH1

Description: This subroutine finds a string in an array of strings and provides the element number in the array of the first matching string. The comparison is case-insensitive.

Argument list:

```
(IFAIL, NSTR, ISTR, ASTR, TSTR)
```

Declarations for argument-list variables:

```
INTEGER,                             INTENT(OUT)   :: IFAIL
INTEGER,                             INTENT(IN)    :: NSTR
INTEGER,                             INTENT(INOUT) :: ISTR
CHARACTER (LEN=*), DIMENSION(NSTR),  INTENT(IN)    :: ASTR
CHARACTER (LEN=*),                   INTENT(INOUT) :: TSTR
```

Explanation of arguments:

IFAIL is a flag that is set to 0 if no errors are encountered, or to the number of errors if any errors are encountered.

NSTR is the dimension of the array of strings to be searched.

ISTR, on invocation, is the element number in the array of strings where the search is to begin. On return, ISTR is the element number containing the matching string.

ASTR is the array of strings to be searched.

TSTR is the string to search for.

Note:

The search is conducted in order of ascending element number unless ISTR equals NSTR on invocation of the subroutine, in which case the search is conducted in order of descending element number.

Data Manipulation Subroutines, Functions, and Generic Interfaces – Matrices

This section documents subroutines, functions, and generic interfaces used primarily to manipulate matrices. The subroutines, functions, and interfaces are arranged alphabetically as follows.

Interface UTL_ARR2CDMATRIX
Subroutine UTL_COMBINESQMATRIX
Subroutine UTL_CONSTRUCTCDMATRIX
Subroutine UTL_CONSTRUCTDMATRIX
Interface UTL_DIAGONAL
Interface UTL_GETCOL
Interface UTL_GETICOL
Interface UTL_GETIROW
Interface UTL_GETPOS
Interface UTL_GETROW
Function UTL_GETVAL
Interface UTL_MATMUL
Interface UTL_MATMULVEC
Subroutine UTL_PREPINVERSE
Subroutine UTL_SVD
Interface UTL_VEC2CDMATRIX

Interface UTL_ARR2CDMATRIX

Description: This generic subroutine interface stores a REAL or DOUBLE PRECISION array in a structure of type CDMATRIX.

Argument list:
```
(NR, NC, ARR, CDM, IDIMOPT, ANAME)
```

Declarations for argument-list variables:
```
INTEGER,                             INTENT(IN)    :: NR
INTEGER,                             INTENT(IN)    :: NC
[various types], DIMENSION(NR,NC),   INTENT(IN)    :: ARR
TYPE (CDMATRIX),                     INTENT(INOUT) :: CDM
INTEGER,          OPTIONAL,          INTENT(IN)    :: IDIMOPT
CHARACTER(LEN=*), OPTIONAL,          INTENT(IN)    :: ANAME
```

Explanation of arguments:
  NR is the number of rows in ARR.
  NC is the number of columns in ARR.
  ARR contains a 2-D array to be stored in the output CDMATRIX structure. ARR may be of type DOUBLE PRECISION or REAL.
  CDM is the CDMATRIX structure to be generated.
  IDIMOPT is the dimension to be used to dimension the DVAL and IPOS component arrays of CDM.
  ANAME is the name to be stored in the ARRAYNAME component of CDM. If ANAME is absent, the ARRAYNAME component is assigned as blank.

Notes:
  1. Only non-zero entries in ARR are stored in CDM.
  2. This subroutine checks the value of each element in ARR to determine the number of non-zero elements (NNZ) that need to be stored in CDM. If NNZ exceeds IDIMOPT (or if IDIMOPT is absent), the DVAL and IPOS are allocated with dimension NNZ.

Subroutine UTL_COMBINESQMATRIX

Description: Use two square matrices (A, B) to generate a third square matrix as a CDMATRIX structure (C) containing the elements of A in the upper left section and the elements of B in the lower right section.

Argument list:
```
(A, B, C, CNAME)
```

Declarations for argument-list variables:
```
TYPE (CDMATRIX),                  INTENT(IN)    :: A
TYPE (CDMATRIX),                  INTENT(IN)    :: B
TYPE (CDMATRIX),                  INTENT(INOUT) :: C
CHARACTER(LEN=*), OPTIONAL, INTENT(IN)          :: CNAME
```

Explanation of arguments:
  A contains a square matrix.
  B contains a square matrix.
  C is populated as a square matrix with the elements of A in the upper left part of the matrix and the elements of B in the lower right part of the matrix. Other elements will be zero.

CNAME is a name which, if present, is assigned to the ARRAYNAME component of the C CDMATRIX structure. ARRAYNAME can accomodate up to 12 characters.

Note:

If A has dimensions j×j and B has dimensions k×k, C will have dimensions (j+k)×(j+k).

Subroutine UTL_CONSTRUCTCDMATRIX

Description: This subroutine generates a properly constructed CDMATRIX structure. Any component array pointers (DVAL, IPOS, and ICOL) that are associated when the subroutine is invoked are deallocated. The component arrays are allocated and populated with zeros. The NNZ component is set to 0.

Argument list:

```
(IDIM, NR, NC, CDM, ANAME)
```

Declarations for argument-list variables:

```
INTEGER,                        INTENT(IN)    :: IDIM
INTEGER,                        INTENT(IN)    :: NR
INTEGER,                        INTENT(IN)    :: NC
TYPE (CDMATRIX),                INTENT(INOUT) :: CDM
CHARACTER(LEN=*), OPTIONAL, INTENT(IN)        :: ANAME
```

Explanation of arguments:

IDIM is the dimension to which the DVAL and IPOS component arrays of the output CDMATRIX structure are to be allocated.

NR is the number of rows that the output CDMATRIX structure will accommodate.

NC is the number of columns that the output CDMATRIX structure will accommodate.

CDM is the output CDMATRIX structure.

ANAME is the name to be assigned to the ARRAYNAME component of the output CDMATRIX structure. If ANAME is absent, the ARRAYNAME component is assigned as blank.

Subroutine UTL_CONSTRUCTDMATRIX

Description: This subroutine generates a properly constructed DMATRIX structure. If the DVAL array pointer component is associated when the subroutine is invoked, the DVAL array is deallocated. The NR and NC components are set equal to the respective arguments, and the double-precision array component (DVAL) is allocated with dimensions (NR, NC). The DVAL array is initialized to 0.0D0.

Argument list:

```
(NR, NC, DM, ANAME)
```

Declarations for argument-list variables:

```
INTEGER,                        INTENT(IN)    :: NR
INTEGER,                        INTENT(IN)    :: NC
TYPE (DMATRIX),                 INTENT(INOUT) :: DM
CHARACTER(LEN=*), OPTIONAL, INTENT(IN)        :: ANAME
```

Explanation of arguments:

NR is the number of rows in the matrix.

NC is the number of columns in the matrix.

DM is the output DMATRIX structure.

ANAME is the name to be assigned to the ARRAYNAME component of the output DMATRIX structure. If ANAME is absent, the ARRAYNAME component is assigned as blank.

Interface UTL_DIAGONAL

Description: This generic function interface populates an array with the diagonal elements extracted from a square matrix stored in a compressed matrix structure.
Argument list:
```
(CDM)
```

Result:
```
ARR
```

Declarations for argument-list and result variables:
```
TYPE (CDMATRIX),  INTENT(IN)           :: CDM
DOUBLE PRECISION, DIMENSION(CDM%NR) :: ARR
```

Explanation of arguments and result:
CDM contains a square matrix.
ARR is populated with diagonal elements in CDM.

Interface UTL_GETCOL

Description: This generic subroutine interface populates a 1-D array with data from one column of a compressed matrix.
Argument list:
```
(CDM, IC, COL, IR1, IR2)
```

Declarations for argument-list variables:
```
TYPE (CDMATRIX),                      INTENT(IN)    :: CDM
INTEGER,                              INTENT(IN)    :: IC
DOUBLE PRECISION, DIMENSION(CDM%NR), INTENT(INOUT) :: COL
INTEGER, OPTIONAL,                    INTENT(IN)    :: IR1
INTEGER, OPTIONAL,                    INTENT(IN)    :: IR2
```

Explanation of arguments:
CDM is the compressed matrix containing data to be used to populate COL.
IC is the number of the column of the true matrix stored in CDM to be copied into COL.
COL is populated with data from column IC of the true matrix stored in CDM.
IR1 is the starting row when a subset of the column data is to be copied.
IR2 is the ending row when a subset of the column data is to be copied.
Notes:
1. Either IR1 and IR2 must both be present, or they must both be absent.
2. If IR1 and IR2 are present, the subsection of array COL designated by COL(IR1:IR2) is populated with corresponding values from column IC of matrix CDM. Values stored in COL outside subsection (IR1:IR2) are not changed.

3. When a 1-dimensional array is to be populated, a single call to UTL_GETCOL is more efficient than populating individual elements of the array using iterative calls to UTL_GETVAL.

Interface UTL_GETICOL

Description: This generic function interface returns the column index for a specified element in a compressed matrix.

Argument list:
```
(CDM, LOC)
```

Result:
```
ICOL
```

Declarations for argument-list and result variables:
```
TYPE (CDMATRIX),  INTENT(IN)  :: CDM
INTEGER,          INTENT(IN)  :: LOC
INTEGER                       :: ICOL
```

Explanation of arguments and result:
   CDM contains a compressed matrix.
   LOC is a location in the IPOS component array of CDM.
   ICOL is the column index associated with element LOC of the IPOS component array of CDM.
Note:
   If LOC is less than 1 or greater than the NNZ component of CDM, ICOL is returned as 0.

Interface UTL_GETIROW

Description: This generic function interface returns the row index for a specified element in a compressed matrix.

Argument list:
```
(CDM, LOC)
```

Result:
```
IROW
```

Declarations for argument-list and result variables:
```
TYPE (CDMATRIX),  INTENT(IN)  :: CDM
INTEGER,          INTENT(IN)  :: LOC
INTEGER                       :: IROW
```

Explanation of arguments and result:
   CDM contains a compressed matrix.
   LOC is a location in the IPOS component array of CDM.
   IROW is the row index associated with element LOC of the IPOS component array of CDM.
Note:
   If LOC is less than 1 or greater than the NNZ component of CDM, IROW is returned as 0.

Appendix D: Utilities (UTL) Module of Chapter 6

Interface UTL_GETPOS

Description: This generic function interface returns as an integer the position in the IPOS array corresponding to the element at row IR, column IC in a compressed matrix structure. This position also applies to the DVAL array of a CDMATRIX structure.

Argument list:
```
(CDM, IR, IC)
```

Result:
```
POS
```

Declarations for argument-list and result variables:
```
TYPE (CDMATRIX), INTENT(IN) :: CDM
INTEGER,         INTENT(IN) :: IR
INTEGER,         INTENT(IN) :: IC
INTEGER,                    :: POS
```

Explanation of arguments and result:
> CDM contains a compressed matrix.
> IR is the row index of interest.
> IC is the column index of interest.
> POS is the position in IPOS corresponding to element (IR, IC) of CDM.

Notes:
> 1. If a value at element (IR, IC) is not stored (indicating that the value at (IR, IC) is zero), 0 is returned.
> 2. If either IR or IC is out of range (less than 1 or greater than CDM%NR or CDM%NC, respectively), -1 is returned.

Interface UTL_GETROW

Description: This generic subroutine interface populates a 1-D array with data from one row of a compressed matrix.

Argument list:
```
(CDM, IR, ROW, IC1, IC2)
```

Declarations for argument-list variables:
```
TYPE (CDMATRIX),                       INTENT(IN)    :: CDM
INTEGER,                               INTENT(IN)    :: IR
DOUBLE PRECISION, DIMENSION(CDM%NC),   INTENT(INOUT) :: ROW
INTEGER, OPTIONAL,                     INTENT(IN)    :: IC1
INTEGER, OPTIONAL,                     INTENT(IN)    :: IC2
```

Explanation of arguments:
> CDM is the compressed matrix containing data to be used to populate ROW.
> IR is the number of the row of the true matrix stored in CDM to be copied into ROW.
> ROW is populated with data from row IR of the true matrix stored in CDM.
> IC1 is the starting column when a subset of the row data is to be copied.
> IC2 is the ending column when a subset of the row data is to be copied.

Notes:
> 1. Either IC1 and IC2 must both be present, or they must both be absent.

175

2. If IC1 and IC2 are present, the subsection of array ROW designated by ROW(IC1:IC2) is populated with corresponding values from row IR of matrix CDM. Values stored in ROW outside subsection (IC1:IC2) are not changed.
3. When a 1-dimensional array is to be populated, a single call to UTL_GETROW is more efficient than populating individual elements of the array using iterative calls to UTL_GETVAL.

Function UTL_GETVAL

Description: This function returns the value in a compressed matrix at a specified row and column.
Argument list:
```
(CDM, IR, IC)
```

Result:
```
V
```

Declarations for argument-list and result variables:
```
TYPE (CDMATRIX),  INTENT(IN)  :: CDM
INTEGER,          INTENT(IN)  :: IR
INTEGER,          INTENT(IN)  :: IC
DOUBLE PRECISION              :: V
```

Explanation of arguments and result:
CDM contains a compressed matrix.
IR is the row index for the matrix element of interest.
IC is the column index for the matrix element of interest.
V is the value at matrix element (IR,IC).

Interface UTL_MATMUL

Description: This generic function interface performs matrix multiplication of an ordinary 2-D array and a compressed matrix.
Argument list:
```
(NR, NC, A, B)
```

Result:
```
C
```

Declarations for argument-list and result variables:
```
INTEGER,                    INTENT(IN)  :: NR
INTEGER,                    INTENT(IN)  :: NC
[various types],            INTENT(IN)  :: A
[various types],            INTENT(IN)  :: B
DOUBLE PRECISION, DIMENSION(:,:) :: C
```

Explanation of arguments and result:
NR is the number of rows in the ordinary 2-D array, which may be either A or B.
NC is the number of columns in the ordinary 2-D array, which may be either A or B.
A is either an ordinary 2-D array of type DOUBLE PRECISION, with dimensions (NR,NC) or a compressed-matrix structure of type CDMATRIX (see Notes).

B is either an ordinary 2-D array of type DOUBLE PRECISION, with dimensions (NR,NC) or a compressed-matrix structure of type CDMATRIX (see Notes).

C is an ordinary 2-D array, with dimensions appropriate for the matrix product [A] x [B].

Notes:

If A is an ordinary array, then B must be a compressed-matrix structure, and the dimensions of C must be (NR,B%NC).

If A is a compressed-matrix structure, then B must be an ordinary 2-D array, and the dimensions of C must be (A%NR,NC).

Interface UTL_MATMULVEC

Description: This generic function interface performs matrix multiplication of a compressed matrix times an ordinary 1-D array, where the 1-D array is assumed to represent a column vector.

Argument list:
```
(NRB, CA, OB)
```

Result:
```
OC
```

Declarations for argument-list and result variables:
```
INTEGER,                                INTENT(IN)  :: NRB
TYPE (CDMATRIX),                        INTENT(IN)  :: CA
DOUBLE PRECISION, DIMENSION(NRB), INTENT(IN)  :: OB
DOUBLE PRECISION, DIMENSION(CA%NR)            :: OC
```

Explanation of arguments and result:

NRB is the dimension of the OB array; it must equal the NC component of the CA structure.

CA is the compressed-matrix structure to be multiplied by OB.

OB contains the column vector used to multiply CA.

OC is assigned as the matrix product [CA] x [OB].

Subroutine UTL_PREPINVERSE

Description: This subroutine prepares a CDMATRIX structure suitable for containing the inverse of the input CDMATRIX structure. The output CDMATRIX structure is constructed using UTL_CONSTRUCTCDMATRIX. The subroutine finds the number of non-zero entries required for the inverse and assigns the NNZ component of the output matrix as this number. The DVAL array is allocated with dimension NNZ and initialized to 0.0D0. The IPOS array is allocated with dimension NNZ and populated correctly for the inverted matrix. The ICOL array is allocated the same as the input CDMATRIX structure and populated correctly for the inverted matrix.

Argument list:
```
(A, AI, AINAME)
```

Declarations for argument-list variables:
```
TYPE (CDMATRIX),                INTENT(IN)    :: A
TYPE (CDMATRIX),                INTENT(INOUT) :: AI
CHARACTER(LEN=*), OPTIONAL, INTENT(IN)    :: AINAME
```

Explanation of arguments:

A is the input CDMATRIX structure.

AI is the output CDMATRIX structure.

AINAME, if present, is assigned to the ARRAYNAME component of the output CDMATRIX structure. If AINAME is absent, the ARRAYNAME component is assigned as blank.

Subroutine UTL_SVD

Description: This subroutine inverts a compressed square, symmetric matrix.
Argument list:
```
(IFAIL, A, AS, DTLA)
```

Declarations for argument-list variables:

```
INTEGER,           INTENT(INOUT) :: IFAIL
TYPE (CDMATRIX),   INTENT(INOUT) :: A
TYPE (CDMATRIX),   INTENT(INOUT) :: AS
DOUBLE PRECISION, INTENT(OUT)    :: DTLA
```

Explanation of arguments:

IFAIL is a flag for error messaging, 0 indicates no problem, 1 indicates failure.

A on invocation contains the matrix to be inverted; on return, A contains the inverted matrix.

AS is populated with the square root of the inverse matrix.

DTLA is the log-determinant of the matrix.

Note:

UTL_SVD invokes subroutines UTL_PREPINVERSE and UTL_DEALLOC.

Interface UTL_VEC2CDMATRIX

Description: This generic subroutine interface converts a 1-D array to a CDMATRIX structure. The non-zero elements in the input array are stored in the diagonal elements of the output matrix. The input array may be either of type REAL or of type DOUBLE PRECISION.
Argument list:
```
(IVEC, VEC, CDM, IDIMOPT, ANAME)
```

Declarations for argument-list variables:
```
INTEGER,                              INTENT(IN)    :: IVEC
[various types], DIMENSION(IVEC),     INTENT(IN)    :: VEC
TYPE (CDMATRIX),                      INTENT(INOUT) :: CDM
INTEGER,                    OPTIONAL, INTENT(IN)    :: IDIMOPT
CHARACTER(LEN=*),           OPTIONAL, INTENT(IN)    :: ANAME
```

Explanation of arguments:

IVEC is the dimension of the VEC array.

VEC is the input 1-D array to be converted to a compressed matrix. VEC can be either of type REAL or of type DOUBLE PRECISION.

CDM is the CDMATRIX structure to be constructed.

IDIMOPT is the minimum dimension for the DVAL and IPOS component arrays of the output CDMATRIX structure. If IDIMOPT is present and larger than the number of nonzero entries in VEC, the DVAL and IPOS arrays are dimensioned with IDIMOPT elements.

ANAME is the value to be assigned to the ARRAYNAME component of the output CDMATRIX structure. If ANAME is absent, the ARRAYNAME component is assigned as blank.

Data Manipulation Subroutines and Generic Interfaces – Input Blocks

This section documents subroutines and generic interfaces used primarily for reading input blocks and storing the results. The subroutines and interfaces are arranged alphabetically as follows.

Subroutine UTL_APPENDLIST
Interface UTL_FILTER
Interface UTL_FILTERLIST
Subroutine UTL_GROUPLIST
Subroutine UTL_MERGELIST

Subroutine UTL_APPENDLIST

Description: Append an input-block data structure (pointed to by LIST2) to another input-block data structure (LIST1) by assigning the NEXTLIST element of the last LLIST structure in LIST1 to point to the head of LIST2.

Argument list:
```
(LIST1, LIST2)
```

Declarations for argument-list variables:
```
TYPE (LLIST), POINTER :: LIST1
TYPE (LLIST), POINTER :: LIST2
```

Explanation of arguments:
LIST1 is an input-block data structure to which another input-block data structure is to be appended.
LIST2 is an input-block data structure.

Interface UTL_FILTER

Description: This generic subroutine interface traverses an input-block data structure and stores a value associated with a given keyword into scalar variable SV. When the type of SV is LOGICAL, and the value associated with the keyword is either "yes," "y," "true," or "t" (comparison is case-insensitive), SV is assigned as .TRUE.; when the value is either "no," "n," "false," or "f," SV is assigned as .FALSE.. For other types of SV, the value associated with the keyword is assigned to SV if the value can be interpreted correctly for the type of SV. A list may contain more than one node (type LNODE) for which the KEYWORD component matches the KEYWORD argument; in this case, the value stored in SV is the last value in the list associated with the matching keyword. If a list does not contain a node for which the KEYWORD component matches the KEYWORD argument, the value of SV is left unchanged. If the list to be traversed is one list in an input-block data structure and is not the first list, the LIST argument can be specified to indicate which list is to be traversed.

Argument list:
```
(IERR, HEAD, IOUT, KEYWORD, SV, LIST)
```

Declarations for argument-list variables:
```
INTEGER,            INTENT(INOUT)  :: IERR
TYPE (LLIST),       POINTER        :: HEAD
INTEGER,            INTENT(IN)     :: IOUT
CHARACTER(LEN=*),   INTENT(IN)     :: KEYWORD
[various types],    INTENT(INOUT)  :: SV
INTEGER, OPTIONAL,  INTENT(IN)     :: LIST
```

Explanation of arguments:
> IERR is a flag that is set to 0 if no errors are encountered, or to the number of errors if any errors are encountered.
> HEAD points to the head of the first list of the nest of lists to be traversed.
> IOUT is the unit number to which error messages, if required, are written.
> KEYWORD is the keyword associated with the data to be stored.
> SV is the scalar variable to be assigned. It may be of type REAL, DOUBLE PRECISION, INTEGER, CHARACTER, or LOGICAL.
> LIST is the number of the list to be traversed (that is, [LIST-1] lists are skipped); the default is 1.

Notes:
1. When HEAD points to an input-block data structure containing a single linked list, the LIST argument should be omitted or specified as 1.
2. When HEAD points to an input-block data structure containing multiple linked lists, the LIST argument must be specified as an integer greater than 1 if the list to be traversed is not the first list in the nest.
3. An input-block data structure containing a single linked list is generated when the KITEM argument of UTL_READBLOCK is specified as an asterisk ("*").
4. An input-block data structure containing multiple linked lists is generated when the KITEM argument of UTL_READBLOCK is not specified as an asterisk and the input block read by UTL_READBLOCK lists multiple occurrences of the keyitem.

Interface UTL_FILTERLIST

Description: This generic subroutine interface traverses an input-block data structure and stores values associated with a given keyword into array ARR. When the type of ARR is LOGICAL, and the value associated with the keyword is either "yes", "y", "true", "t", or "on" (comparison is case-insensitive), the corresponding element of ARR is assigned as .TRUE.; when the value is either "no", "n", "false", "f", or "off", the corresponding element of ARR is assigned as .FALSE.. For other types of ARR, the value associated with the keyword is assigned to the corresponding element of ARR if the value can be interpreted correctly for the type of ARR. A list may contain more than one node (type LNODE) for which the KEYWORD component matches the KEYWORD argument; in this case, the value stored in ARR is the last value in the list associated with the matching keyword. If a list does not contain a node for which the KEYWORD component matches the KEYWORD argument, the contents of the ARR element corresponding to that list is left unchanged. The "Example Application: GROUP_EXAMPLE" section of Chapter 18 illustrates the use of UTL_FILTERLIST.

Argument list:
```
(HEAD, IOUT, KEYWORD, NDIM, IERR, ARR, MORE, IARRSTART, LISTSTART)
```

Declarations for argument-list variables:
```
TYPE (LLIST),                       POINTER          :: HEAD
INTEGER,                            INTENT(IN)       :: IOUT
CHARACTER(LEN=*),                   INTENT(IN)       :: KEYWORD
INTEGER,                            INTENT(IN)       :: NDIM
INTEGER,                            INTENT(INOUT)    :: IERR
[various types], DIMENSION(NDIM),   INTENT(INOUT)    :: ARR
INTEGER,                            INTENT(OUT)      :: MORE
INTEGER, OPTIONAL,                  INTENT(IN)       :: IARRSTART
INTEGER, OPTIONAL,                  INTENT(IN)       :: LISTSTART
```

Explanation of arguments:

HEAD points to the head of the first list of the nest of lists to be traversed.

IOUT is the unit number to which error messages, if required, are written.

KEYWORD is the keyword associated with the data to be stored.

NDIM is the declared dimension of ARR.

IERR is a flag that is set to 0 if no errors are encountered, or to the number of errors if any errors are encountered.

ARR is the 1-dimensional array to be populated. It may be of type REAL, DOUBLE PRECISION, INTEGER, CHARACTER, or LOGICAL, and it is declared with dimension NDIM.

MORE is the number of lists left to process after last element in array ARR has been assigned.

IARRSTART is the element number in array ARR at which to start assigning values from the input-block data structure; the default is 1.

LISTSTART is the number of the list at which to start processing values (that is, [LISTSTART-1] lists are skipped before data transfer starts); the default is 1.

Notes:

1. Elements of ARR are assigned starting at element IARRSTART, or at element 1 if IARRSTART is absent or is less than 1.

2. Data transfer starts at list number LISTSTART, or at list number 1 if LISTSTART is absent or is less than 1.

3. Data transfer proceeds from the input-block data structure to the array with the element number in the array incremented as each list is processed. (Note that the array element number is incremented even if a list does not contain an entry for which the keyword matches the KEYWORD of the argument list.) Data transfer continues until the last list is processed or until the last element in ARR is reached, whichever occurs first. If additional lists remain unprocessed after the last element in ARR is assigned, the number of lists left unprocessed is assigned to MORE.

Subroutine UTL_GROUPLIST

Description: This subroutine inserts linked-list nodes (LNODE structures) associated with a group into the individual lists in an input-block data structure immediately after the head of individual lists for which the group name matches. The effect of this insertion is to provide keywords and values defined by group to members of each group as defaults. The "Example Application: GROUP_EXAMPLE" section of Chapter 18 illustrates the use of UTL_GROUPLIST.

Argument list:
```
(DEFGROUP, GRPHEAD, IOUT, LSTHEAD, NGROUPS, NLISTS)
```

Declarations for argument-list variables:
```
CHARACTER(LEN=12),  INTENT(IN) :: DEFGROUP
TYPE (LLIST),       POINTER    :: GRPHEAD
INTEGER,            INTENT(IN) :: IOUT
TYPE (LLIST),       POINTER    :: LSTHEAD
INTEGER,            INTENT(IN) :: NGROUPS
INTEGER,            INTENT(IN) :: NLISTS
```

Explanation of arguments:
   DEFGROUP is the default group name for individual lists (see note).
   GRPHEAD points to the input-block data structure containing group information.
   IOUT is the unit number to which error messages, if any, are written.
   LSTHEAD points to the input-block data structure containing information on individual
      members of groups.
   NGROUPS is the number of groups.
   NLISTS is the number of lists of individual members.
Note:
   If the input-block structure pointed to by GRPHEAD includes a group with GROUPNAME that
      matches DEFGROUP, any list in the input-block structure pointed to by LSTHEAD that
      does not have GROUPNAME defined will be matched with group DEFGROUP, and
      insertion of group data will take place accordingly. As a result, keywords and values
      defined for group DEFGROUP in the GRPHEAD structure will be the default values for
      lists in the LSTHEAD structure that have no explicitly defined group.

Subroutine UTL_MERGELIST

Description: This subroutine merges two input-block data structures by adding linked lists
   included in HEAD2 to the tails of corresponding linked lists included in HEAD1. A list in
   HEAD2 corresponds to a list in HEAD1 if the contents of the VALUE components of the first
   node in each list match each other. The comparison is case-insensitive. If no match is found,
   the case is flagged as an error.
Argument list:
```
(BLOCKLABEL1, BLOCKLABEL2, HEAD1, HEAD2, IOUT)
```

Declarations for argument-list variables:
```
CHARACTER(LEN=*),  INTENT(IN) :: BLOCKLABEL1
CHARACTER(LEN=*),  INTENT(IN) :: BLOCKLABEL2
TYPE (LLIST),      POINTER    :: HEAD1
TYPE (LLIST),      POINTER    :: HEAD2
INTEGER,           INTENT(IN) :: IOUT
```

Explanation of arguments:
   BLOCKLABEL1 is the blocklabel for the input block (Chapter 3) which was read to create the
      input-block data structure pointed to by HEAD1.
   BLOCKLABEL2 is the blocklabel for the input block which was read to create the input-block
      data structure pointed to by HEAD2.
   HEAD1 points to the input-block data structure to which nodes are to be added.
   HEAD2 points to the input-block data structure containing nodes to be added to HEAD1.
   IOUT is the unit number to which error messages, if any, are written.

Data Manipulation Subroutines, Functions, and Generic Interface – Other

This section documents subroutines, functions, and one generic interface used primarily to copy data, perform calculations, or convert data between data types. The subroutines, functions, and interface are arranged alphabetically as follows.

Interface ASSIGNMENT(=)
Subroutine UTL_COVMAT
Function UTL_ELAPSED_TIME
Subroutine UTL_ENDTIME
Function UTL_GETUNIT
Function UTL_NEXTUNIT
Function UTL_SUBSTITUTE
Subroutine UTL_SYSTEM

Interface ASSIGNMENT(=)

Description: This generic interface overloads the "=" assignment operator to enable assignment statements to be used with specific derived data types. This subroutine need not be called explicitly; it is invoked implicitly when an assignment statement of a supported form is executed in a subprogram that USEs the Utilities Module. Supported forms include (1) assignment of a CDMATRIX structure equal to another CDMATRIX structure; or (2) assignment of a 2-D DOUBLE PRECISION array equal to a CDMATRIX structure. The name of this interface is required by the Fortran-90 standard (American National Standards Institute, 1992) to be ASSIGNMENT(=); therefore, the JUPITER API naming convention cannot be applied.

Argument list:
```
(DOUT, DIN)
```

Declarations for argument-list variables:
```
[various types], INTENT(INOUT) :: DOUT
[various types], INTENT(IN)    :: DIN
```

Explanation of arguments:
DOUT is the output data array or structure of the type of a supported form. In an assignment statement, it is the entity to the left of the "=" sign. In the initial release of the JUPITER API, the only supported data types for DOUT are CDMATRIX and DOUBLE PRECISION; future versions of the API may support additional options.
DIN is the input data structure of the type of a supported form. In an assignment statement, it is the entity to the right of the "=" sign. In the initial release of the JUPITER API, the only supported data type for DIN is CDMATRIX; future versions of the API may support additional options.

Notes:
1. Example usage in code:
Form 1: "A=B" where A and B are structures of type CDMATRIX.
Form 2: "A2D=A" where A2D is a 2-D DOUBLE PRECISION array and A is a structure of type CDMATRIX; note that A2D must have been declared or allocated with first and second dimensions equal to A%NR and A%NC, respectively.
2. To assign a CDMATRIX structure equal to a 2-D array, use the UTL_ARR2CDMATRIX subroutine.

3. To assign a CDMATRIX structure equal to a 1-D array, use the UTL_VEC2CDMATRIX subroutine.

Subroutine UTL_COVMAT

Description: This subroutine builds a combined variance-covariance matrix from values obtained from a set of variance-covariance matrices for groups of observations, predictions, or prior-information equations (for correlated groups in the set), or from an array of variances (for uncorrelated groups in the set). Each member (each observation, each prediction, or each prior-information equation) is assumed to belong to a group. The resulting matrix is a CDMATRIX structure containing an NMEM by NMEM matrix, where NMEM is the number of members.

Argument list:
```
(IOUT, NCOVMAT, NMEM, NGPS, COVMATARR, COVMATNAM, GROUP, GROUPNAM, MEMNAM,
VARIANCE, WTCORR, CDCOVMAT)
```

Declarations for argument-list variables:
```
INTEGER,                                    INTENT(IN)    :: IOUT
INTEGER,                                    INTENT(IN)    :: NCOVMAT
INTEGER,                                    INTENT(IN)    :: NMEM
INTEGER,                                    INTENT(IN)    :: NGPS
TYPE (CDMATRIX), DIMENSION(NCOVMAT),        INTENT(IN)    :: COVMATARR
CHARACTER(LEN=12), DIMENSION(NGPS),         INTENT(IN)    :: COVMATNAM
CHARACTER(LEN=12), DIMENSION(NMEM),         INTENT(IN)    :: GROUP
CHARACTER(LEN=12), DIMENSION(NGPS),         INTENT(IN)    :: GROUPNAM
CHARACTER(LEN=LENDNAM), DIMENSION(NMEM),    INTENT(IN)    :: MEMNAM
DOUBLE PRECISION, DIMENSION(NMEM),          INTENT(IN)    :: VARIANCE
LOGICAL, DIMENSION(NMEM),                   INTENT(IN)    :: WTCORR
TYPE (CDMATRIX),                            INTENT(INOUT) :: CDCOVMAT
```

Explanation of arguments:
IOUT is the unit number associated with the file to which output, if any, is to be written.
NCOVMAT is the number of variance-covariance matrices stored in the COVMATARR array.
NMEM is the dimension of the GROUP, MEMNAM, VARIANCE, and WTCORR arrays; it is the number of members in all groups in the set of groups of a given type (that is, all observation groups, all prediction groups, or all prior-information equation groups).
NGPS is the number of groups in the set of groups.
COVMATARR contains, as a minimum, a variance-covariance matrix for each group that is listed by name in the GROUP array and is identified as being correlated by TRUE values in the WTCORR array. COVMATARR may contain additional matrices that are not listed in the GROUP array.
COVMATNAM contains the name of the variance-covariance matrix associated with each group, according to the order of groups in the GROUPNAM array. If a group's members are uncorrelated, the value of COVMATNAM should be blank.
GROUP contains the name of the group to which each member belongs.
GROUPNAM contains one entry, populated with the group name, for each distinct group listed in the GROUP array.
MEMNAM contains the names of all members.
VARIANCE contains the variance associated with each member. The contents of the VARIANCE array are used only for members of uncorrelated groups.

WTCORR is TRUE for members of correlated groups, and FALSE for members of uncorrelated groups.

CDCOVMAT is populated with the NMEM by NMEM variance-covariance matrix for all members.

Function UTL_ELAPSED_TIME

Description: This function gets the current time and calculates an elapsed time, relative to an earlier time provided as an argument.
Argument list:
```
(IBDT)
```

Result:
```
ELTIME
```

Declarations for argument-list and result variables:
```
INTEGER, DIMENSION(8), INTENT(IN) :: IBDT
DOUBLE PRECISION                  :: ELTIME
```

Explanation of argument and result:
IBDT contains the date and time data generated from an earlier invocation of the DATE_AND_TIME intrinsic Fortran subroutine.
ELTIME is the elapsed time, in seconds.

Subroutine UTL_ENDTIME

Description: This subroutine gets the current time and calculates an elapsed time, relative to an earlier time provided as an argument. The elapsed time is written to the screen and is optionally written to an output file unit.
Argument list:
```
(IBDT, IOUT)
```

Declarations for argument-list variables:
```
INTEGER, DIMENSION(8), INTENT(IN) :: IBDT
INTEGER,               INTENT(IN) :: IOUT
```

Explanation of arguments:
IBDT contains the date and time data generated from an earlier invocation of the DATE_AND_TIME intrinsic Fortran procedure.
IOUT is the unit number associated with the file to which output is to be written. If IOUT is less than 1, output is not written to a file unit.

Function UTL_GETUNIT

Description: This function returns the first unused file unit number in the range IFIRST through MAXUNIT.
Argument list:
```
(IFIRST, MAXUNIT)
```

Result:
```
KUNIT
```

Declarations for argument-list and result variables:
```
INTEGER, INTENT(IN) :: IFIRST
INTEGER, INTENT(IN) :: MAXUNIT
INTEGER            :: KUNIT
```

Explanation of arguments and result:
IFIRST is the unit number to be checked.
MAXUNIT is the last unit number to be checked.
KUNIT is the smallest number in the range IFIRST through MAXUNIT that is unused as a unit number.

Notes:
1. MAXUNIT should be greater than or equal to IFIRST, and both should be positive.
2. A unit number (I) is considered unused if the OPENED specifier of the Fortran INQUIRE statement is returned as FALSE when INQUIRE is invoked with the specification UNIT=I.

Function UTL_NEXTUNIT

Description: This function returns the first unused unit number in the range 10 to 1,000.
Argument list: Empty
Result:
```
NEXTUNIT
```
Declaration for result variable:
```
INTEGER :: NEXTUNIT
```

Function UTL_SUBSTITUTE

Description: This function populates an array of dimension NPT with parameter values using current values in the PVAL array, but with substitution of values for all adjustable parameters provided in the PARVALSET array.
Argument list:
```
(NPE, NPT, IPTR, PARVALSET, PVAL)
```

Result:
```
PVTMP
```

Declarations for argument-list and result variables:
```
INTEGER,                           INTENT(IN) :: NPE
INTEGER,                           INTENT(IN) :: NPT
INTEGER, DIMENSION(NPE),           INTENT(IN) :: IPTR
DOUBLE PRECISION, DIMENSION(NPE),  INTENT(IN) :: PARVALSET
DOUBLE PRECISION, DIMENSION(NPT),  INTENT(IN) :: PVAL
DOUBLE PRECISION, DIMENSION(NPT)              :: PVTMP
```

Explanation of arguments:
NPE is the number of adjustable parameters.
NPT is the number of parameters.

IPTR contains, for each adjustable parameter, the position (element number) in the PVAL array corresponding to the adjustable parameter.

PARVALSET contains values of adjustable parameters, which may differ from the current parameter values, for example, for perturbation of a parameter.

PVAL contains the current (unperturbed) parameter values.

PVTMP is populated with values from PVAL with substitution of values in the PARVALSET array for all adjustable parameters.

Subroutine UTL_SYSTEM

Description: This subroutine submits a command to the operating system for execution.
Argument list:
`(COMMAND)`

Declaration for argument-list variable:
`CHARACTER(LEN=*), INTENT(IN) :: COMMAND`

Explanation of arguments:
COMMAND is an operating-system command to be executed by the operating system.

Public Error-Processing Subroutines

This section documents three subroutines used primarily to process errors. The subroutines are arranged alphabetically as follows.
Subroutine UTL_STOP
Subroutine UTL_SUBERROR
Subroutine UTL_VERSION_CHECK

Subroutine UTL_STOP

Description: This subroutine writes, either to the screen or to unit IOUT, the nonblank contents of the AMESSAGE variable of the Global Data Module and the nonblank contents of the STOPMESS argument, if present, and then stops program execution.
Argument list:
`(STOPMESS, IOUT)`

Declarations for argument-list variables:
```
CHARACTER(LEN=*), OPTIONAL, INTENT(IN) :: STOPMESS
INTEGER,          OPTIONAL, INTENT(IN) :: IOUT
```

Explanation of arguments:
STOPMESS is a message to be written to the screen or to IOUT prior to stopping program execution.
IOUT, if present, is the unit number associated with the file to which output is to be written. If IOUT is absent and STOPMESS is non-blank, output is written to the screen.
Note:
UTL_STOP calls UTL_WRITE_MESSAGE. If AMESSAGE of the Global Data Module is non-blank, UTL_WRITE_MESSAGE writes AMESSAGE to the screen or, if IOUT is present, to IOUT.

Appendix D: Utilities (UTL) Module of Chapter 6

Subroutine UTL_SUBERROR

Description: This subroutine writes the name of a subroutine passed to it as the argument and stops program execution.
Argument list:
```
(SUBNAME)
```

Declarations for argument-list variables:
```
CHARACTER (LEN=*) :: SUBNAME
```

Explanation of arguments:
   SUBNAME is the name of the subroutine from which the subroutine is called.

Subroutine UTL_VERSION_CHECK

Description: This subroutine determines if the current version of the API (VERSIONID of the Global Data Module) is as new or newer than the version required by the calling program unit.
Argument list:
```
(MINVERSION, ISTAT)
```

Declarations for argument-list variables:
```
CHARACTER(LEN=*),  INTENT(IN)  :: MINVERSION
INTEGER, OPTIONAL, INTENT(OUT) :: ISTAT
```

Explanation of arguments:
   MINVERSION is a string containing the version information, using the "integer.integer.integer" format defined for VERSIONID (Appendix C).
   ISTAT, if present, is assigned the value 0 if VERSIONID is as new or newer than MINVERSION or the value -1 if VERSIONID is older than MINVERSION.
Notes:
   1. If MINVERSION is newer than VERSIONID and ISTAT is absent, an error message is written to the screen and execution stops.
   2. In comparing MINVERSION against VERSIONID to determine which is older, first the major release numbers (the first integer) are compared. If the integer from MINVERSION is larger than the integer from VERSIONID, then MINVERSION is newer than VERSIONID; if the integer from MINVERSION is smaller than the integer from VERSIONID, then MINVERSION is older than VERSIONID; if they are the same value, then a similar comparison is made on the next pair of integers (minor release number). If both the major and minor release numbers are the same, then a similar comparison is done on the subrelease numbers.

Appendix E: Basic (BAS) Module of Chapter 7

Public Data

Declaration of variables:
```
CHARACTER (LEN=12), ALLOCATABLE, DIMENSION(:) :: COMMANDID
CHARACTER (LEN=12), ALLOCATABLE, DIMENSION(:) :: COMPURPOSE
TYPE (CDMATRIX),    ALLOCATABLE, DIMENSION(:) :: COVMATARR
CHARACTER(LEN=MAX_STRING_LEN)                 :: DERIV_INTERFACE
LOGICAL                                       :: FORSENS
TYPE (LLIST), POINTER                         :: LLPTRMATFIL
TYPE (LLIST), POINTER                         :: LLPTRMODCOM
TYPE (LLIST), POINTER                         :: LLPTROPT
CHARACTER (LEN=MAX_STRING_LEN), ALLOCATABLE, DIMENSION(:) :: MODCOMLINE
INTEGER                                       :: NCOMLINES
```

Explanation of variables:

COMMANDID is allocated with dimension NCOMLINES and populated with model-command identifiers in subroutine BAS_INI_MODELEXEC. The command identifiers are values associated with the keyword COMMANDID in the Model_Command_Lines input block.

COMPURPOSE is allocated with dimension NCOMLINES and populated with command purposes in subroutine BAS_INI_MODELEXEC. The command purposes are values associated with the keyword PURPOSE in the Model_Command_Lines input block.

COVMATARR contains all group variance-covariance matrices.

DERIV_INTERFACE is the value associated with the keyword DERIVATIVES_INTERFACE, if present, in the Options input block. DERIV_INTERFACE is assigned in subroutine BAS_INI_GETOPTIONS and is the name of a derivatives interface file (Appendix A), or blank if not assigned in an Options input block.

FORSENS is true if any command in the Model_Command_Lines input block is specified with PURPOSE="FORWARD&DER" and false otherwise. FORSENS is assigned in subroutine BAS_INI_MODELEXEC.

LLPTRMATFIL is a pointer to an input-block data structure of matrix-file names.

LLPTRMODCOM is a pointer to an input-block data structure of model-command information.

LLPTROPT is a pointer to an input-block data structure of options.

MODCOMLINE is allocated with dimension NCOMLINES and populated with model command lines in subroutine BAS_INI_MODELEXEC.

NCOMLINES is the number of model command lines listed in the Model_Command_Lines input block. NCOMLINES is assigned in subroutine BAS_INI_MODELEXEC.

Private Data

Declaration of variables:
```
INTEGER, PARAMETER                                    :: NMODCOMCOLS = 1
CHARACTER(LEN=40), DIMENSION(NMODCOMCOLS), TARGET :: MODCOMCOL =    &
     (/ 'COMMAND  ','PURPOSE  ','COMMANDID' /)
CHARACTER(LEN=40), DIMENSION(0) :: NOCOL
```

Explanation of variables:

NMODCOMCOLS is the default number of columns expected in the Model_Command_Lines input block when the TABLE blockformat is used and the COLUMNLABELS option is not specified (see Table 7-4).

MODCOMCOL contains the default column headings for the Model_Command_Lines input block when the TABLE blockformat is used and the COLUMNLABELS option is not specified (see Table 7-4).

NOCOL is a zero-dimension array used in calls to UTL_READBLOCK (Appendix D) to read input blocks for which no default column order is defined (Chapter 3).

Public Subroutines – Initialize (INI)

This section documents three subroutines used primarily to initialize arrays and variables. The subroutines are arranged alphabetically as follows.

Subroutine BAS_INI_COVMAT
Subroutine BAS_INI_GETOPTIONS
Subroutine BAS_INI_MODELEXEC

Subroutine BAS_INI_COVMAT

Description: This subroutine calls UTL_READBLOCK to read an input block labeled "Matrix_Files" (Chapters 7 and 16) when the NCOVMAT argument is greater than zero; the REQUIRED argument of UTL_READBLOCK is true. The subroutine allocates the COVMATARR array with a dimension of NCOVMAT, and populates the array with matrices read from the files listed in the input block.

Argument list:

```
(INUNIT, IOUT, NCOVMAT)
```

Declarations for argument-list variables:

```
INTEGER, INTENT(IN) :: INUNIT
INTEGER, INTENT(IN) :: IOUT
INTEGER, INTENT(IN) :: NCOVMAT
```

Explanation of arguments:

INUNIT is the unit number of the file from which an input block labeled "Options" is to be read.

IOUT is the unit number associated with the file to which output is to be written.

NCOVMAT is the number of variance-covariance matrices to be populated.

Notes:

1. If NCOVMAT is less than zero, control returns to the calling program unit with no action taken.

2. The matrices are read by calls to UTL_READMATRIX. Accordingly, the matrices to be read can be provided in either the Complete Matrix or the Compressed Matrix format (Appendix A).

Subroutine BAS_INI_GETOPTIONS

Description: This subroutine calls UTL_READBLOCK to read an input block labeled "Options"; the REQUIRED argument of UTL_READBLOCK is false, so the Options input block is optional. If an Options input block is found, UTL_FILTER is called to assign IVERB of the Global Data Module with the value associated with the keyword VERBOSE (if present), and to assign DERIV_INTERFACE with the value associated with the keyword DERIVATIVES_INTERFACE (if present).

Argument list:
```
(INUNIT, IOUT)
```

Declarations for argument-list variables:
```
INTEGER, INTENT(IN) :: INUNIT
INTEGER, INTENT(IN) :: IOUT
```

Explanation of arguments:
>INUNIT is the unit number of the file from which an input block labeled "Options" is to be read.
>IOUT is the unit number associated with the file to which output is to be written.

Note:
>The call to UTL_READBLOCK in BAS_INI_GETOPTIONS, if an "Options" input block is found, causes the pointer LLPTROPT to be associated and to point to an input-block data structure. In the call to UTL_READBLOCK, the KITEM argument is defined as "*"; as a result, LLPTROPT points to an input-block data structure containing a single linked list. See the "Public Data" section of this Appendix for more information.

Subroutine BAS_INI_MODELEXEC

Description: This subroutine calls UTL_READBLOCK to read an input block labeled "Model_Command_Lines" (Chapters 7 and 16); the REQUIRED argument of UTL_READBLOCK is true, so the "Model_Command_Lines" input block is required. The subroutine then stores and echoes model command lines. Command lines are identified by the keyword "COMMAND", which is also the keyitem. The keywords PURPOSE and COMMANDID also are recognized and associated values are stored by BAS_INI_MODELEXEC. Model commands are stored in the MODCOMLINE array, which is declared as public in the Basic Module.

Argument list:
```
(INUNIT, IOUT)
```

Declarations for argument-list variables:
```
INTEGER, INTENT(IN) :: INUNIT
INTEGER, INTENT(IN) :: IOUT
```

Explanation of arguments:
>INUNIT is the unit number of the file from which an input block labeled "OPTIONS" is to be read.
>IOUT is the unit number associated with the file to which output is to be written.

Note:
>If model-calculated sensitivities are to be supported, before calling BAS_INI_MODELEXEC, BAS_INI_GETOPTIONS should be called to assign a value to DERIV_INTERFACE.

Appendix E: Basic (BAS) Module of Chapter 7

Public Subroutine – Generate Parameter Values (GEN)

This section documents one subroutine used primarily in parameter-value generation.

Subroutine BAS_GEN

Description: This subroutine adjusts current parameter values in PVAL to ensure that all parameter values are equal to the values that will be written to model-input files.
Argument list:
```
(NOPNT, NPT, NW, PRECPR, PVAL)
```

Declarations for argument-list variables:
```
INTEGER,                              INTENT(IN)    :: NOPNT
INTEGER,                              INTENT(IN)    :: NPT
INTEGER,             DIMENSION(NPT),  INTENT(IN)    :: NW
INTEGER,                              INTENT(IN)    :: PRECPR
DOUBLE PRECISION,    DIMENSION(NPT),  INTENT(INOUT) :: PVAL
```

Explanation of arguments:
   NOPNT is a flag indicating the decimal point protocol for writing parameter values.
      NOPNT=0 indicates that a decimal point should be included when writing parameter values to a model-input file, and NOPNT=1 indicates that a decimal point should not be included.
   NPT is the number of parameters.
   NW contains the minimum number of characters available for writing each parameter to model-input files.
   PRECPR is a flag indicating the precision protocol for writing parameter values. Valid values for PRECPR and their meanings are:
      PRECPR = 0 – values will be written in single precision, with exponent in "E" format when required.
      PRECPR = 1 – values will be written in double precision, with exponent in "E" format when required.
      PRECPR = 2 – values will be written in double precision, with exponent in "D" format when required.
   PVAL contains a set of parameter values. These values are adjusted, if necessary, to equal the values that will be written to model-input files.

Public Subroutines – Execute Process Model (EXE)

This section documents the following two subroutines use dot execute process models.
Subroutine BAS_EXE
Subroutine BAS_EXE_SELECT

Subroutine BAS_EXE

Description: This subroutine executes the model command stored in position ICOMMAND in the MODCOMLINE array (stored by subroutine BAS_INI_MODELEXEC) as an operating-system command.
Argument list:
```
(ICOMMAND, IOUT, KRUN, MRUNS)
```

Declarations for argument-list variables:
```
INTEGER,           INTENT(IN) :: ICOMMAND
INTEGER,           INTENT(IN) :: IOUT
INTEGER, OPTIONAL, INTENT(IN) :: KRUN
INTEGER, OPTIONAL, INTENT(IN) :: MRUNS
```

Explanation of arguments:
ICOMMAND is the command number (position in list of commands stored in MODCOMLINE).
IOUT is the unit number associated with the file to which output is to be written.
KRUN is a model run number.
MRUNS is the number of model runs in a set of runs.
Notes:
1. This subroutine makes use of a compiler-specific extension to Fortran-90 to execute a system command. Editing this subroutine may be required if the extension used in this subroutine is not supported by your compiler.
2. If IVERB of the Global Data Module is greater than 1, messages may be written to unit IOUT or to the screen indicating, at appropriate times, that the specified command has been invoked and that it has finished. If IOUT is greater than zero, the messages are written to unit IOUT. If IOUT is less than zero, the messages are written to the screen. If IOUT equals zero, no messages are written.
3. If KRUN is present and MRUNS is absent, the text "(run KRUN)" is appended to the message indicating that the command has been invoked.
4. If both KRUN and MRUNS are present, the text "(run KRUN of MRUNS)" is appended to the message indicating that the command has been invoked.

Subroutine BAS_EXE_SELECT

Description: This subroutine determines the number of a model command to be executed. Two options are supported. If JOB is "FORWARD", the command number of the first command for which PURPOSE is "FORWARD" is returned in ICOMMAND. IF JOB is "FORWARD&SENS", the command number of the first command for which PURPOSE is "FORWARD&DER" is returned.

Argument list:
```
(IOUT, JOB, ICOMMAND, COMMAND)
```

Declarations for argument-list variables:
```
INTEGER,                               INTENT(IN)  :: IOUT
CHARACTER(LEN=*),                      INTENT(IN)  :: JOB
INTEGER,                               INTENT(OUT) :: ICOMMAND
CHARACTER(LEN=MAX_STRING_LEN), OPTIONAL, INTENT(OUT) :: COMMAND
```

Explanation of arguments:
IOUT is the unit number associated with the file to which output is to be written.
JOB is either "FORWARD" or "FORWARD&SENS".
ICOMMAND is the command number (position in list of commands stored in MODCOMLINE).
COMMAND is the command text associated with command ICOMMAND.

Public Subroutine – Cleanup (CLN)

Subroutine BAS_CLN

Description: This subroutine deallocates all allocatable arrays that are stored in the Basic Module.
Argument list: Empty

Appendix F: Model Input-Output (MIO) Module of Chapter 8

Public Data

Declaration of variables:
```
TYPE (LLIST), POINTER :: LLPTRMODIN
TYPE (LLIST), POINTER :: LLPTRMODOUT
```

Explanation of variables:
> LLPTRMODIN points to an input-block data structure populated from an input block labeled "Model_Input_Files" by a call to UTL_READBLOCK from MIO_INI_INPUTFILES.
> LLPTRMODOUT points to an input-block data structure populated from an input block labeled "Model_Output_Files" by a call to UTL_READBLOCK from MIO_INI_OUTPUTFILES.

Public Subroutines – Initialize (INI)

This section documents subroutines used primarily for initialization. The subroutines are presented alphabetically, as follows.
> Subroutine MIO_INI_ALLOC
> Subroutine MIO_INI_ARRAYS
> Subroutine MIO_INI_DIMENSION
> Subroutine MIO_INI_INPUTFILES
> Subroutine MIO_INI_INPUTFILES_RUNNER
> Subroutine MIO_INI_INSTRUCT1
> Subroutine MIO_INI_INSTRUCT2
> Subroutine MIO_INI_INSTRUCTALLOC
> Subroutine MIO_INI_INSTRUCT_RUNNER
> Subroutine MIO_INI_OUTPUTFILES
> Subroutine MIO_INI_OUTPUTFILES_RUNNER
> Subroutine MIO_INI_TEMPLATE

Subroutine MIO_INI_ALLOC

Description: This subroutine initializes variables of the Model Input-Output Module.
Argument list:
```
(IFAIL, NPT)
```

Declarations for argument-list variables:
```
INTEGER, INTENT(OUT) :: IFAIL
INTEGER, INTENT(IN)  :: NPT
```

Explanation of arguments:
> IFAIL is an error flag. IFAIL is set to 0 if no error is encountered. IFAIL is set to 1 if an error is encountered.
> NPT is the number of parameters.

Subroutine MIO_INI_ARRAYS

Description: This subroutine populates argument-list arrays with values from arrays of the Model
   Input-Output Module.

Argument list:
```
(IFAIL, LCIS, NINSTRUCT, NMIFILE, NMOFILE, CATGOR, CINSTSET, INSTRUCTFILE,
LOCINS, MIFILE, MOFILE, MRKDL, TFILE)
```

Declarations for argument-list variables:
```
INTEGER,                                           INTENT(OUT) :: IFAIL
INTEGER,                                           INTENT(IN)  :: LCIS
INTEGER,                                           INTENT(IN)  :: NINSTRUCT
INTEGER,                                           INTENT(IN)  :: NMIFILE
INTEGER,                                           INTENT(IN)  :: NMOFILE
CHARACTER(LEN=6),              DIMENSION(NMOFILE), INTENT(OUT) :: CATGOR
CHARACTER(LEN=1),              DIMENSION(LCIS),    INTENT(OUT) :: CINSTSET
CHARACTER(LEN=MAX_STRING_LEN), DIMENSION(NMOFILE), INTENT(OUT) :: INSTRUCTFILE
INTEGER,                       DIMENSION(NINSTRUCT),INTENT(OUT) :: LOCINS
CHARACTER(LEN=MAX_STRING_LEN), DIMENSION(NMIFILE), INTENT(OUT) :: MIFILE
CHARACTER(LEN=MAX_STRING_LEN), DIMENSION(NMOFILE), INTENT(OUT) :: MOFILE
CHARACTER(LEN=1),              DIMENSION(NMOFILE), INTENT(OUT) :: MRKDL
CHARACTER(LEN=MAX_STRING_LEN), DIMENSION(NMIFILE), INTENT(OUT) :: TFILE
```

Explanation of arguments:
   IFAIL is an error flag. Any value other that 0 indicates an error has occurred.
   LCIS is the dimension (module variable LENCIS) of the CISET module array, as provided by
      the MIO_INI_DIMENSION subroutine.
   NINSTRUCT is the dimension of the LCINS module array, as provided by the
      MIO_INI_DIMENSION subroutine.
   NMIFILE is the dimension of the MODINFILE, TEMPFILE, and PARDEL module arrays, as
      provided by the MIO_INI_DIMENSION subroutine.
   NMOFILE is the dimension of the MODOUTFILE, INSFILE, MRKDEL, and CATEGORY
      module arrays, as provided by the MIO_INI_DIMENSION subroutine.
   CATGOR is populated with the contents of the CATEGORY module array.
   CINSTSET is populated with the contents of the CISET module array.
   INSTRUCTFILE is populated with the contents of the INSFILE module array.
   LOCINS is populated with the contents of the LCINS module array.
   MIFILE is populated with the contents of the MODINFILE module array.
   MOFILE is populated with the contents of the MODOUTFILE module array.
   MRKDL is populated with the contents of the MRKDEL module array.
   TFILE is populated with the contents of the TEMPFILE module array.

Appendix F: Model Input-Output (MIO) Module of Chapter 8

Subroutine MIO_INI_DIMENSION

Description: This subroutine returns dimensions of selected arrays of the Model Input-Output Module.

Argument list:
```
(LCIS, NINSTRUCT, NMIFILE, NMOFILE, NUMLADV)
```

Declarations for argument-list variables:
```
INTEGER, INTENT(OUT) :: LCIS
INTEGER, INTENT(OUT) :: NINSTRUCT
INTEGER, INTENT(OUT) :: NMIFILE
INTEGER, INTENT(OUT) :: NMOFILE
INTEGER, INTENT(OUT) :: NUMLADV
```

Explanation of arguments:

LCIS is assigned the value of the module variable LENCIS, which is the dimension of the CISET module array.

NINSTRUCT is assigned the value of the module variable NINSTR, which is the dimension of the LCINS module array.

NMIFILE is assigned the value of the module variable NUMINFILE, which is the number of model-input files and the dimension of the MODINFILE, TEMPFILE, and PARDEL module arrays.

NMOFILE is assigned the value of the module variable NUMOUTFILE, which is number of model-output files and the dimension of the MODOUTFILE, INSFILE, MRKDEL, and CATEGORY module arrays.

NUMLADV is assigned the value of the module variable NUML, which is the dimension of the LADV module array.

Subroutine MIO_INI_INPUTFILES

Description: This subroutine calls UTL_READBLOCK to read an input block labeled "Model_Input_Files" (Table 16-17); the REQUIRED argument of UTL_READBLOCK is true, so the "Model_Input_Files" input block is required. This input block is expected to contain the columns or keywords MODINFILE and TEMPLATEFILE. When keywords are used, MODINFILE is the keyitem. Each MODINFILE entry is used to designate the name of a model-input file, and each corresponding TEMPLATEFILE entry is used to designate the name of a corresponding template file (See "Template Files" section of Appendix A). The module variable NUMINFILE is assigned, the MODINFILE and TEMPFILE arrays are allocated, and these arrays are populated with data read from the "Model_Input_Files" input block.

Argument list:
```
(INUNIT, IOUT)
```

Declarations for argument-list variables:
```
INTEGER, INTENT(IN) :: INUNIT
INTEGER, INTENT(IN) :: IOUT
```

Explanation of arguments:

INUNIT is the unit number associated with the file from which the "Model_Input_Files" input block is to be read.

IOUT is the unit number associated with the file to which output is to be written.

Subroutine MIO_INI_INPUTFILES_RUNNER

Description: This subroutine assigns the module variable NUMINFILE; allocates module arrays MODINFILE, TEMPFILE, and PARDEL; and populates arrays MODINFILE and TEMPFILE.

Argument list:
```
(IFAIL, NMIFILE, MIFILE, TFILE)
```

Declarations for argument-list variables:
```
INTEGER,                                               INTENT(OUT) :: IFAIL
INTEGER,                                               INTENT(IN)  :: NMIFILE
CHARACTER (LEN=MAX_STRING_LEN), DIMENSION(NMIFILE), INTENT(IN)  :: MIFILE
CHARACTER (LEN=MAX_STRING_LEN), DIMENSION(NMIFILE), INTENT(IN)  :: TFILE
```

Explanation of arguments:
> IFAIL is an error flag. Any value other than 0 indicates an error has occurred.
> NMIFILE is the number of model-input files; its value is assigned to module variable NUMINFILE.
> MIFILE contains the names of the model-input files. It is used to populate module array MODINFILE.
> TFILE contain the names of the template files used in preparing model-input files. It is used to populate module array TEMPFILE.

Subroutine MIO_INI_INSTRUCT1

Description: This subroutine reads all extraction-instruction files to determine memory needs and assign values to dimensioning variables.

Argument list:
```
(IFAIL)
```

Declarations for argument-list variables:
```
INTEGER, INTENT(OUT) :: IFAIL
```

Explanation of arguments:
> IFAIL is an error flag. IFAIL is set to 0 if no error is encountered. IFAIL is set to 1 if an error is encountered.

Notes:
1. The INSFILE array must have been populated with the names of instruction files prior to calling MIO_INI_INSTRUCT1.
2. The MODOUTFILE array must have been populated with the names of model-output files from which simulated-equivalent or prediction values are to be read prior to calling MIO_INI_INSTRUCT1.
3. Subroutine MIO_INI_OUTPUTFILES can be called prior to calling MIO_INI_INSTRUCT1 to allocate and populate the INSFILE and MODOUTFILE arrays.
4. The marker delimiters read from the extraction-instruction files are stored in the MRKDEL array.

Subroutine MIO_INI_INSTRUCT2

Description: This subroutine reads and stores extraction instructions.
Argument list:
```
(IFAIL, NCATS, MCVCAT)
```

Declarations for argument-list variables:
```
INTEGER,                                INTENT(OUT) :: IFAIL
INTEGER,                                INTENT(IN)  :: NCATS
CHARACTER (LEN=6), DIMENSION(NCATS), INTENT(IN)  :: MCVCAT
```

Explanation of arguments:
> IFAIL is an error flag. IFAIL is set to 0 if no error is encountered. IFAIL is set to 1 if an error is encountered.
> NCATS is the number of model-calculated-value categories supported by the application.
> MCVCAT contains the keyword associated with each model-calculated-value category supported by the application.

Notes:
1. The INSFILE array must have been populated with the names of instruction files prior to calling MIO_INI_INSTRUCT2.
2. The MODOUTFILE array must have been populated with the names of model-output files from which simulated-equivalent or prediction values are to be read prior to calling MIO_INI_INSTRUCT2.
3. The MRKDEL array must have been populated with the marker delimiter read from each extraction-instruction file prior to calling MIO_INI_INSTRUCT2.
4. Subroutine MIO_INI_OUTPUTFILES can be called prior to calling MIO_INI_INSTRUCT2 to allocate and populate the INSFILE and MODOUTFILE arrays and to allocate the MRKDEL array.
5. Subroutine MIO_INI_INSTRUCT1 can be called prior to calling MIO_INI_INSTRUCT2 to populate the MRKDEL array.

Subroutine MIO_INI_INSTRUCTALLOC

Description: This subroutine allocates module arrays CISET, LADV, LCILNS, INS1, INS2, and INS3, which are used for storing extraction instructions.
Argument list:
```
(NDEP, IFAIL)
```

Declarations for argument-list variables:
```
INTEGER, INTENT(IN)  :: NDEP
INTEGER, INTENT(OUT) :: IFAIL
```

Explanation of arguments:
> NDEP is the number of dependent variables for which extraction instructions are to be stored
> IFAIL is an error flag. IFAIL is set to 0 if no error is encountered. IFAIL is set to 1 if an error is encountered.

Appendix F: Model Input-Output (MIO) Module of Chapter 8

Subroutine MIO_INI_INSTRUCT_RUNNER

Description: This subroutine assigns module variables LENCIS, NINSTR, and NUML; calls
    MIO_INI_INSTRUCTALLOC to allocate module arrays for storing extraction instructions; and
    populates module arrays CISET and LCINS from arrays in argument list.

Argument list:
```
(IFAIL, LCIS, NINSTRUCT, NUMLADV, NVEXT, CINSTSET, LOCINS)
```

Declarations for argument-list variables:
```
INTEGER,                               INTENT(OUT) :: IFAIL
INTEGER,                               INTENT(IN)  :: LCIS
INTEGER,                               INTENT(IN)  :: NINSTRUCT
INTEGER,                               INTENT(IN)  :: NUMLADV
INTEGER,                               INTENT(IN)  :: NVEXT
CHARACTER(LEN=1), DIMENSION(LCIS),     INTENT(IN)  :: CINSTSET
INTEGER,          DIMENSION(NINSTRUCT), INTENT(IN)  :: LOCINS
```

Explanation of arguments:
    IFAIL is an error flag. IFAIL is set to 0 if no error is encountered. IFAIL is set to 1 if an error
        is encountered.
    LCIS is assigned to module variable LENCIS.
    NINSTRUCT is assigned to module variable NINSTR.
    NUMLADV is assigned to module variable NUML.
    NVEXT is the number of dependent values to be extracted from model-output files.
    CINSTSET is used to populate the CISET module array.
    LOCINS is used to populate the LCINS module array.

Subroutine MIO_INI_OUTPUTFILES

Description: This subroutine allocates, initializes, and populates arrays with information related to
    model-output files and extraction of model-calculated values. It calls UTL_READBLOCK to
    read an input block labeled "Model_Output_Files" (Table 16-1); the REQUIRED argument of
    UTL_READBLOCK is true, so the "Model_Output_Files" input block is required. This input
    block is expected to contain the columns or keywords MODOUTFILE, INSTRUCTIONFILE,
    and CATEGORY. When keywords are used, MODOUTFILE is the keyitem. Each
    MODOUTFILE entry is used to designate the name of a model-output file, and each
    corresponding INSTRUCTIONFILE entry is used to designate the name of a corresponding
    extraction-instruction file (See "Template Files" section of Appendix A). The CATEGORY
    entry is used to designate the category of model-calculated values to be extracted from each
    model-output file, where the permitted categories are application-defined.

Argument list:
```
(INUNIT, IOUT)
```

Declarations for argument-list variables:
```
INTEGER, INTENT(IN) :: INUNIT
INTEGER, INTENT(IN) :: IOUT
```

Explanation of arguments:
    INUNIT is the unit number associated with the file from which the "Model_Output_Files"
        input block is to be read.

IOUT is the unit number associated with the file to which output is to be written.

Subroutine MIO_INI_OUTPUTFILES_RUNNER

Description: This subroutine allocates and populates module arrays needed for extracting model-calculated dependent values from model-output files. It assigns the module variable NUMOUTFILE; allocates module arrays MODOUTFILE, INSFILE, MRKDEL, and CATEGORY; and populates these arrays with data from arrays in the argument list.

Argument list:
```
(IFAIL, NMOFILE, CATGOR, INSTRUCTFILE, MOFILE, MRKDL)
```

Declarations for argument-list variables:
```
INTEGER,                                                 INTENT(OUT) :: IFAIL
INTEGER,                                                 INTENT(IN)  :: NMOFILE
CHARACTER (LEN=6),                  DIMENSION(NMOFILE), INTENT(IN)  :: CATGOR
CHARACTER (LEN=MAX_STRING_LEN), DIMENSION(NMOFILE), INTENT(IN)  :: INSTRUCTFILE
CHARACTER (LEN=MAX_STRING_LEN), DIMENSION(NMOFILE), INTENT(IN)  :: MOFILE
CHARACTER (LEN=1),                  DIMENSION(NMOFILE), INTENT(IN)  :: MRKDL
```

Explanation of arguments:
>IFAIL an error flag. Any value other than 0 indicates an error has occurred.
>NMOFILE is the number of model-output files.
>CATGOR is used to populate the CATEGORY module array.
>INSTRUCTFILE is used to populate the INSFILE module array.
>MOFILE contains the names of the model-output files and is used to populate the MODOUTFILE module array.
>MRKDL contains the marker delimiters and is used to populate the MRKDEL module array.

Subroutine MIO_INI_TEMPLATE

Description: This subroutine does some checking of template files and finds the smallest character width to which each parameter will be written.

Argument list:
```
(IFAIL, NPT, APAR, NW)
```

Declarations for argument-list variables:
```
INTEGER,                                 INTENT(OUT) :: IFAIL
INTEGER,                                 INTENT(IN)  :: NPT
CHARACTER (LEN=*), DIMENSION(NPT), INTENT(IN)  :: APAR
INTEGER, DIMENSION(NPT),              INTENT(OUT) :: NW
```

Explanation of arguments:
>IFAIL is an error flag. IFAIL is set to 0 if no error is encountered. IFAIL is set to 1 if an error is encountered.
>NPT is the number of parameters.
>APAR contains the parameter names.
>NW contains the number of characters available for writing each parameter to model-input files.

Public Subroutines – Adapt Parameter Values (ADA)

This section documents two subroutines:
Subroutine MIO_ADA_MODINFILE_CHANGE
Subroutine MIO_ADA_WRITEFILES

Subroutine MIO_ADA_MODINFILE_CHANGE

Description: This subroutine stores a file name in a selected element of the MODINFILE array.
Argument list:
```
(IFAIL, IFILE, INFILENAME)
```

Declarations for argument-list variables:
```
INTEGER,          INTENT(OUT) :: IFAIL
INTEGER,          INTENT(IN)  :: IFILE
CHARACTER(LEN=*), INTENT(IN)  :: INFILENAME
```

Explanation of arguments:
   IFAIL is an error flag. IFAIL is set to 0 if no error is encountered. IFAIL is set to 1 if an error
      is encountered.
   IFILE is the element number in MODINFILE where the file name is to be stored.
   INFILENAME is the file name to be stored.
Note:
   INFILENAME may be either a file name, absolute pathname, or relative pathname.

Subroutine MIO_ADA_WRITEFILES

Description: This subroutine writes model-input files, using parameter values supplied in the
   argument list.
Argument list:
```
(IFAIL, NPT, APAR, NOPNT, NW, PRECPR, PVAL)
```

Declarations for argument-list variables:
```
INTEGER,                            INTENT(OUT)   :: IFAIL
INTEGER,                            INTENT(IN)    :: NPT
CHARACTER (LEN=*), DIMENSION(NPT),  INTENT(IN)    :: APAR
INTEGER,                            INTENT(IN)    :: NOPNT
INTEGER, DIMENSION(NPT),            INTENT(IN)    :: NW
INTEGER,                            INTENT(IN)    :: PRECPR
DOUBLE PRECISION, DIMENSION(NPT),   INTENT(INOUT) :: PVAL
```

Explanation of arguments:
   IFAIL is an error flag. IFAIL is set to 0 if no error is encountered. IFAIL is set to 1 if an error
      is encountered.
   NPT is the number of parameters.
   APAR contains the parameter names.
   NOPNT is a flag indicating the decimal point protocol for writing parameter values.
      NOPNT=0 indicates that a decimal point should be included when writing parameter values
      to a model-input file, and NOPNT=1 indicates that a decimal point should not be included.

NW contains the number of characters available for writing each parameter to model-input files.

PRECPR is a flag indicating the precision protocol for writing parameter values. Valid values for PRECPR and their meanings are:

PRECPR = 0 – values will be written in single precision, with exponent in "E" format when required.

PRECPR = 1 – values will be written in double precision, with exponent in "E" format when required.

PRECPR = 2 – values will be written in double precision, with exponent in "D" format when required.

PVAL contains the parameter values to be used in preparing the model-input files.

Public Subroutines – Extract Model-Calculated Values (EXT)

This section documents two subroutines:
Subroutine MIO_EXT
Subroutine MIO_EXT_MODOUTFILE_CHANGE

Subroutine MIO_EXT

Description: This subroutine reads model-output files and extracts model-calculated values by using a set of extraction instructions.

Argument list:
```
(IFAIL, IOUT, NCATS, NVEXT, EXTNAM, MCVUSE, EXTVAL, INSTRUCTION)
```

Declarations for argument-list variables:
```
INTEGER,                                   INTENT(OUT) :: IFAIL
INTEGER,                                   INTENT(IN)  :: IOUT
INTEGER,                                   INTENT(IN)  :: NCATS
INTEGER,                                   INTENT(IN)  :: NVEXT
CHARACTER (LEN=*), DIMENSION(NVEXT),       INTENT(IN)  :: EXTNAM
LOGICAL,           DIMENSION(NCATS),       INTENT(IN)  :: MCVUSE
DOUBLE PRECISION,  DIMENSION(NVEXT),       INTENT(OUT) :: EXTVAL
CHARACTER (LEN=*),                         INTENT(OUT) :: INSTRUCTION
```

Explanation of arguments:

IFAIL is an error flag. IFAIL is set to 0 if no error is encountered. IFAIL is set to 1 if an error is encountered.

IOUT is the unit number associated with the file to which output is to be written. If IOUT is less than 1, output is not written to a file unit.

NCATS is the number of model-calculated-value categories supported by the application.

NVEXT is the number of values to be extracted.

EXTNAM contains names with which extracted values are to be associated.

MCVUSE indicates, for each model-calculated-value category supported by the application, whether extractions should be performed on model-output files associated with the category.

EXTVAL contains the extracted values.

INSTRUCTION contains the instruction in error when IFAIL is returned as 1.

Notes:
1. If an error condition occurs and the INSTRUCTION variable is not blank, then this variable contains the offending instruction. The contents of this variable should be reproduced with the error message.
2. Each model-output file has an associated application-supported category, as defined by the contents of the CATEGORY module array. Each supported category must correspond to an element of MCVUSE, which indicates if that category of model-calculated values is to be extracted. Model-calculated values are extracted only from model-output files for which the element of MCVUSE corresponding to the category assigned that model-output file is true.
3. If IVERB of the Global Data Module is greater than 3, extracted values for dependents are written to unit IOUT.

Subroutine MIO_EXT_MODOUTFILE_CHANGE

Description: This subroutine stores a file name in a selected element of the MODOUTFILE array.
Argument list:
```
(IFAIL, IFILE, OUTFILENAME)
```

Declarations for argument-list variables:
```
INTEGER,           INTENT(OUT) :: IFAIL
INTEGER,           INTENT(IN)  :: IFILE
CHARACTER(LEN=*),  INTENT(IN)  :: OUTFILENAME
```

Explanation of arguments:
IFAIL is an error flag. IFAIL is set to 0 if no error is encountered. IFAIL is set to 1 if an error is encountered.
IFILE is the element number in MODOUTFILE where the file name is to be stored.
OUTFILENAME is the file name to be stored.
Note:
OUTFILENAME may be either a file name, absolute pathname, or relative pathname.

Public Subroutine – Cleanup (CLN)

Subroutine MIO_CLN_DEALLOC

Description: This subroutine deallocates memory used by the Model Input-Output Module.
Argument list: Empty

Subroutine MIO_CLN_OUTFILES

Description: This subroutine deletes all model-output files listed in the MODOUTFILE array.
Argument list:
```
(IFAIL)
```

Declarations for argument-list variables:
```
INTEGER,                                INTENT(OUT) :: IFAIL
```

Appendix F: Model Input-Output (MIO) Module of Chapter 8

Explanation of arguments:

IFAIL is an error flag. IFAIL is set to 0 if no error is encountered. IFAIL is set to 1 if an error is encountered.

Appendix G: Equation (EQN) Module of Chapter 9

Derived Data Type

The Equation Module defines one derived data type, called "COMPRESSED_EQUATION". One structure of type COMPRESSED_EQUATION is designed to store one equation. The definition is:

```
TYPE COMPRESSED_EQUATION
  CHARACTER (LEN=12)            :: NAME
  INTEGER                       :: NUMTERM
  INTEGER                       :: LASTINDEX
  CHARACTER (LEN=1), POINTER    :: ASTRING(:)
  INTEGER, POINTER              :: PTERM(:)
END TYPE COMPRESSED_EQUATION
```

where components are defined as follows:
  NAME is the equation name.
  NUMTERM is the number of terms in the equation.
  LASTINDEX is highest used index of ASTRING array.
  ASTRING contains the equation.
  PTERM contains the location of terms of equation in ASTRING array.

Public Data

No public data is made available by the Equation Module.

Private Data

```
INTEGER, PARAMETER                        :: LOPER=8
INTEGER, PARAMETER                        :: MAXTERM=500
INTEGER, PARAMETER                        :: MAXEXP=10
INTEGER, PARAMETER                        :: NFUNCT=19
INTEGER, PARAMETER                        :: NOPER=15
INTEGER                                   :: NUMEQN=0
INTEGER                                   :: IORDER(MAXTERM)
INTEGER                                   :: MTYPE(MAXTERM)
INTEGER                                   :: MTYPE1(MAXTERM)
INTEGER                                   :: MAX_EQN_LENGTH=MAX_STRING_LEN
INTEGER                                   :: LASTJPAR=1
INTEGER                                   :: NTERM
CHARACTER (LEN=MAX_STRING_LEN)            :: BTEXT
CHARACTER (LEN=25)                        :: ATERM(MAXTERM)
CHARACTER (LEN=25)                        :: BTERM(MAXTERM)
CHARACTER (LEN=25)                        :: CTERM(MAXTERM)
CHARACTER, DIMENSION(LOPER)               :: LOPERAT
CHARACTER (LEN=5), DIMENSION(LOPER)       :: LOPERATFULL
CHARACTER, DIMENSION(NOPER)               :: OPERAT
DOUBLE PRECISION                          :: DTERM(MAXTERM)
DOUBLE PRECISION                          :: ETERM(MAXTERM)
LOGICAL                                   :: LTERM(MAXTERM)
LOGICAL                                   :: OTERM(MAXTERM)
TYPE (COMPRESSED_EQUATION),ALLOCATABLE    :: COMPEQN(:)
CHARACTER (LEN=6), DIMENSION(NFUNCT)      :: FUNCT
```

```
DATA FUNCT /'abs   ','acos  ','asin  ','atan  ','cos   ','cosh  ',   &
    'exp   ','log   ','log10 ','sin   ','sinh  ','sqrt  ','tan   ',   &
    'tanh  ','neg   ','pos   ','min   ','max   ','mod   '/
```

Explanation of variables:

LOPER is number of logical operators.

MAXTERM is maximum number of elements in an equation.

MAXEXP is maximum number of exponents in an equation.

NFUNCT is number of functions.

NOPER is number of operators.

NUMEQN is number of equations.

IORDER is an integer work array.

MTYPE is an integer work array.

MTYPE1 is an integer work array.

MAX_EQN_LENGTH is maximum length of equation text.

LASTJPAR is the number of the last parameter found in an equation.

NTERM is number of active terms in ATERM.

BTEXT is temporary text storage.

ATERM is temporary equation storage.

BTERM is temporary equation storage.

CTERM is temporary equation storage.

LOPERAT contains logical operators.

LOPERATFULL contains unabbreviated logical operators.

OPERAT contains all operators, including arithmetic and logical operators.

DTERM contains numbers in equation.

ETERM contains numbers in equation.

LTERM contains logical terms in equation.

OTERM contains logical terms in equation.

COMPEQN contains compressed equations.

FUNCT contains function names.

Public Subroutines

This section documents subroutines used primarily to enable equations. The subroutines are listed as follows.

Subroutine EQN_INI

Subroutine EQN_INI_INSTALL

Subroutine EQN_CLN

Subroutine EQN_EVALUATE

Subroutine EQN_GETNAME

Subroutine EQN_LINEAR_COEFFS

Subroutine EQN_INI

Description: Subroutine EQN_INI is used to initialize the Equation Module. It must be called prior to calling any other subroutines of this module.

Argument list:

```
(IFAIL, NUM_EQN)
```

Appendix G: Equation (EQN) Module of Chapter 9

Declarations for argument-list variables:
```
INTEGER, INTENT(OUT)    :: IFAIL
INTEGER, INTENT(IN)     :: NUM_EQN
```

Explanation of arguments:

IFAIL signals an error condition. If it is returned as non-zero, an error condition has occurred. An error message will be available in the AMESSAGE string made available by the Global Data Module.

NUM_EQN is the number of equations requiring installation. It is important to note that only equations which are to be stored and repeatedly evaluated by the Equation Module need to be considered when counting equations to evaluate NUM_EQN. Linear prior-information equations should not be counted in the evaluation of NUM_EQN as they are neither stored nor evaluated — only parsed. If only linear prior-information equations are to be used, then NUM_EQN should be supplied as zero.

Subroutine EQN_INI_INSTALL

Description:

Subroutine EQN_INI_INSTALL installs equations that are later to be evaluated. These equations are first parsed and then stored in compressed form for later access. The user must also assign each equation a number; this number must be between 1 and NUM_EQN inclusive (NUM_EQN is the total number of evaluatable equations previously supplied as the second argument to subroutine EQN_INI). When subroutine EQN_EVALUATE is later used to evaluate an equation, the equation is referred to by this number (though error messages will cite the actual equation name). Note that equation numbers must not be duplicated; each equation must have a unique integer identifier associated with it.

An equation is supplied as a line of text. As presently programmed, this line must be no more than 2,000 characters in length and contain no more that 500 separate entities. (An "entity" is an operator, parameter name, bracket, function name, or variable.) In parsing the equation, it is assumed that any entity that is not recognizable as an operator, function, or number is a variable. Variable names, however, are not checked against a list of allowed names on equation installation; such checking takes place only when an equation is evaluated in subroutine EQN_EVALUATE. This separation of installation and variable verification allows flexibility in the handling of equations. For example, some equations may cite variables that are parameters, while others may cite variables that are observations, while still others may cite control variables such as iteration number, objective function, and so forth, which the main program makes available as a basis for decision making. All such equations can be installed irrespective of the variables that they use. When an equation is later evaluated, however, an array of variable names and values pertinent to that equation must then be supplied.

Linear prior-information equations should not be installed using subroutine EQN_INI_INSTALL. These equations are processed only by subroutine EQN_LINEAR_COEFFS.

Argument list:
```
(IFAIL, IEQN, EQUATION_NAME, ATEXT)
```

Declarations for argument-list variables:
```
INTEGER, INTENT(OUT)           :: IFAIL
INTEGER, INTENT(IN)            :: IEQN
CHARACTER (LEN=*), INTENT(IN)  :: EQUATION_NAME
CHARACTER (LEN=*), INTENT(IN)  :: ATEXT
```

Explanation of arguments:

IFAIL signals an error condition. If it is returned as non-zero, an error condition has occurred. An error message will be available in the AMESSAGE string made available by the Global Data Module.

IEQN is the equation number. This must be assigned by the user. It must be between 1 and the total number of evaluatable equations. It must not be the same as that assigned to an equation which has already been installed.

EQUATION_NAME is the name of the equation. This must be 12 characters or less in length. It is used by subroutines of the Equation Module to refer to equations for which an error condition is reported.

ATEXT is a text string containing the equation. No "=" symbol is allowed in this string; the text must consist entirely of the evaluatable part of the equation. Maximum equation length is 2,000 characters.

Subroutine EQN_CLN

Description: Subroutine EQN_CLN deallocates memory used by the Equation Module. It should be called prior to termination of any program that USEs the Equation Module.
Argument list: Empty

Subroutine EQN_EVALUATE

Description: Subroutine EQN_EVALUATE evaluates an equation that has previously been parsed and stored by subroutine EQN_INI_INSTALL. The outcome of the equation can be a numerical or logical value, depending on its contents.
Argument list:

```
(IFAIL, IEQN, NVAR, AVAR, RVAR, ITYPE, RVAL, LVAL, UNASSIGNED, ANAME)
```

Declarations for argument-list variables:

```
INTEGER,                                INTENT(OUT) :: IFAIL
INTEGER,                                INTENT(IN)  :: IEQN
INTEGER,                                INTENT(IN)  :: NVAR
CHARACTER (LEN=*), DIMENSION(NVAR), INTENT(IN)  :: AVAR
DOUBLE PRECISION,   DIMENSION(NVAR), INTENT(IN)  :: RVAR
INTEGER,                                INTENT(OUT) :: ITYPE
DOUBLE PRECISION,                       INTENT(OUT) :: RVAL
LOGICAL,                                INTENT(OUT) :: LVAL
DOUBLE PRECISION,          OPTIONAL, INTENT(IN)  :: UNASSIGNED
CHARACTER (LEN=*),         OPTIONAL, INTENT(IN)  :: ANAME
```

Explanation of arguments:

IFAIL signals an error condition. If it is returned as non-zero, an error condition has occurred. An error message will be available in the AMESSAGE string made available by the Global Data Module.

IEQN is the equation number – the same number that was used to identify the equation when the EQN_INI_INSTALL subroutine was used to install this equation.

NVAR is the number of variables.

AVAR contains a list of NVAR variable names (each of 12 characters or less in length). **Note that variable names must be supplied in lower case.**

RVAR contains values for the variables contained in AVAR. These values are substituted into the equation in place of corresponding variable names when the equation is evaluated.

ITYPE is returned as 0 if the equation has a numerical outcome, and as 1 if it has a logical outcome. In the former case, the equation outcome is returned as RVAL; in the latter case, it is returned as LVAL.

RVAL is the numerical value of an equation that has a numerical outcome.

LVAL is the logical value of an equation that has a logical outcome.

UNASSIGNED is a value which, if present, is compared with values in RVAR before they are used. If a required value in RVAR equals UNASSIGNED, the value in RVAR is assumed to be invalid, an error message is assigned to the AMESSAGE variable of the Global Data Module, IFAIL is set to 1, and control returns to the calling program unit.

ANAME is an equation name. If IEQN equals zero, ANAME must be present.

Notes:
1. The equation to be evaluated must be identified either by number (IEQN) or by name (ANAME). If ANAME is present and IEQN is greater than zero, IEQN is used to identify the equation and ANAME is ignored.
2. By initializing elements of the RVAR array and UNASSIGNED with a number not expected to be validly used in RVAR, for example -10^{35} or BIGDOUBLE of the Global Data Module, failure to assign a valid value to a required element of RVAR can be caught as an error.

Subroutine EQN_GETNAME

Description: Subroutine EQN_GETNAME retrieves the name of an installed equation, given its number.

Argument list:
```
(IFAIL, IEQN, EQUATION_NAME)
```

Declarations for argument-list variables:
```
INTEGER,            INTENT(OUT) :: IFAIL
INTEGER,            INTENT(IN)  :: IEQN
CHARACTER (LEN=*), INTENT(OUT) :: EQUATION_NAME
```

Explanation of arguments:
IFAIL signals an error condition. If it is returned as non-zero then an error condition has occurred. An error message will be available in the AMESSAGE string available from the Global Data Module.

IEQN is the equation number — the same number that was used to identify the equation when the EQN_INI_INSTALL subroutine was used to install this equation.

EQUATION_NAME is the name of the equation — the same name that was installed using EQN_INI_INSTALL.

Subroutine EQN_LINEAR_COEFFS

Description: Subroutine EQN_LINEAR_COEFFS is used for the processing of linear prior information. It does not actually install linear prior-information equations; hence, such equations are not made available for evaluation at a later date. Rather, this subroutine evaluates coefficients of variables cited in the equation for insertion into a row of the Jacobian matrix.

Appendix G: Equation (EQN) Module of Chapter 9

There are very strict rules governing the legality of linear prior-information equations. If any of these rules are violated, EQN_LINEAR_COEFFS will cease processing and return execution to the main program with an appropriate error message. These rules are now outlined.

1. A linear prior-information equation must specify a linear relationship with respect to any adjustable parameters, or their logs, that are cited within it. If a cited parameter is log-transformed for the purpose of parameter estimation, then the log (to base 10 or e) of that parameter must be cited in the equation; the untransformed parameter must not be cited. If a cited parameter is not log-transformed for the purpose of parameter estimation, then the log of that parameter must not be cited in a linear prior-information equation; only the native parameter can be cited. Failure to obey this rule, as well as the presence of any other sources of parameter nonlinearity, will be detected by subroutine EQN_LINEAR_COEFFS and reported.

2. A parameter that is tied to another parameter must not be cited in a linear prior-information equation.

3. A fixed parameter can be cited in a linear prior-information equation. Obviously, there is no place in the Jacobian matrix for such a parameter. However, as EQN_LINEAR_COEFFS calculates the value of the equation when all adjustable parameters (or their logs) are zero while fixed parameter are assigned their fixed values, the calling program has the option of adjusting the "observed value" of a linear prior-information equation to account for the presence of such fixed parameters within it. This functionality removes the requirement that prior-information equations be altered as a user fixes and (or) unfixes parameters during successive runs of a JUPITER program.

4. A linear prior-information equation must not use brackets, unless they are used in conjunction with the log() or log10() function. Nor can it use the "/" or "**" operators. Nor can it use any functions.

5. A linear prior-information equation must cite at least one adjustable parameter.

The following are examples of legal linear prior-information equations.
```
par1
par1*2
2*par1 + 5 * log(par2)
5.43 + 13.4*log10(par2) -9*par3+par4
```

The following linear prior-information equations are illegal.
```
par1*log( par2)
3.2*sin(1.0)*par4
4.0/2.0 * par1
```

In the above examples it is assumed that *par1* to *par4* are all adjustable, and that only *par2* is log-transformed.

Argument list:
```
(IFAIL, EQUATION_NAME, ATEXT, NVAR, NPE, AVAR, ITRANS, RVAR, CONST_TERM, WORK,
XROW)
```

Declarations for argument-list variables:
```
INTEGER, INTENT(OUT)            ::  IFAIL
CHARACTER (LEN=*), INTENT(IN)   ::  EQUATION_NAME
CHARACTER (LEN=*), INTENT(IN)   ::  ATEXT
INTEGER, INTENT(IN)             ::  NVAR
INTEGER, INTENT(IN)             ::  NPE
CHARACTER (LEN=*), INTENT(IN)   ::  AVAR(NVAR)
INTEGER, INTENT(IN)             ::  ITRANS(NVAR)
```

```
DOUBLE PRECISION, INTENT(IN)    :: RVAR(NVAR)
DOUBLE PRECISION, INTENT(OUT)   :: CONST_TERM
DOUBLE PRECISION, INTENT(OUT)   :: WORK(NVAR)
DOUBLE PRECISION, INTENT(OUT)   :: XROW(NPE)
```

Explanation of arguments:

IFAIL signals an error condition. If it is returned as non-zero then an error condition has occurred. An error message will be available in the AMESSAGE string available from the Global Data Module. In this error message the equation will be identified by its name (12 characters or less) supplied by the calling program through the EQUATION_NAME variable.

EQUATION_NAME is the equation name.

ATEXT contains the text of the linear prior-information equation. It must contain only the part of the equation that can be evaluated; it must not contain an "=" sign.

NVAR is the number of parameters for which elements are populated in the AVAR, ITRANS and RVAR arrays.

NPE is the number of parameters that are adjustable through the parameter estimation process. This is also equal to the number of columns of the Jacobian matrix.

AVAR is an array of NVAR parameter names; corresponding parameter values need to be supplied in the RVAR array. Parameter names need to be lower case.

ITRANS contains the transformation state for each parameter in the AVAR array. An ITRANS value of 0 indicates that the pertinent parameter is untransformed, whereas a value of 1 indicates that it is log-transformed. A value of $-10,000$ (that is, negative 10,000) indicates that the parameter is fixed (at its corresponding RVAR value); any other negative value indicates that the parameter is tied, the index of the parent parameter being the negative of the value of ITRANS for that parameter. Exactly NPE elements of ITRANS must have a value of either 0 or 1.

RVAR contains the values of parameters identified in the AVAR array. Note that the only use subroutine EQN_LINEAR_COEFFS makes of the elements of the RVAR array is in the calculation of the constant term for the equation; fixed parameters contribute to that term.

CONST_TERM is the value taken by the linear prior-information equation when all adjustable parameters (or their logs) are assigned a value of zero. Thus it takes into account constant terms used in the equation as well as non-zero values of fixed parameters.

WORK is an array used by EQN_LINEAR_COEFFS only for internal processing.

XROW is populated with values that can be used to directly fill a row of the Jacobian matrix. It contains linear coefficients of adjustable parameters (that is, parameters for which ITRANS is either 0 or 1) in order of their appearance in the AVAR array. However, because the coefficients of only adjustable parameters are calculated, parameters whose ITRANS value is less than zero are omitted from this array. Elements pertaining to parameters not cited in a linear prior-information equation are assigned a value of zero in the XROW array. It is important to note that where a parameter is log-transformed (that is, ITRANS for that parameter is supplied as 1), the pertinent element of the XROW array contains the coefficient of the log to base 10 of that parameter irrespective of whether the supplied equation uses the log of the parameter to base 10, or to base e.

Appendix H: Dependents (DEP) Module of Chapter 10

Public Data

Declaration of variables:
```
CHARACTER(LEN=12), ALLOCATABLE, DIMENSION(:)       :: GROUP
CHARACTER(LEN=12), ALLOCATABLE, DIMENSION(:)       :: GROUPNAM
CHARACTER(LEN=12), ALLOCATABLE, DIMENSION(:)       :: GROUPNOBS
INTEGER,           ALLOCATABLE, DIMENSION(:)       :: IGPTYPE
TYPE (LLIST), POINTER                              :: LLPTRDEP
TYPE (LLIST), POINTER                              :: LLPTRDEPGP
DOUBLE PRECISION,  ALLOCATABLE, DIMENSION(:)       :: MODEXTVAL
DOUBLE PRECISION,  ALLOCATABLE, DIMENSION(:)       :: MODDERVAL
DOUBLE PRECISION,  ALLOCATABLE, DIMENSION(:)       :: NONDETVAL
INTEGER                                            :: NPREDGPS
INTEGER                                            :: NTOTOBS
INTEGER                                            :: NTOTPRED
INTEGER                                            :: NUSEDEP
INTEGER                                            :: NUSEOBS
INTEGER                                            :: NUSEPRED
INTEGER,           ALLOCATABLE, DIMENSION(:)       :: OBSDEROBSNUM
CHARACTER(LEN=LENDNAM), ALLOCATABLE, DIMENSION(:)  :: OBSEXTNAM
INTEGER,           ALLOCATABLE, DIMENSION(:)       :: OBSEXTOBSNUM
DOUBLE PRECISION,  ALLOCATABLE, DIMENSION(:)       :: OBSVAL
LOGICAL,           ALLOCATABLE, DIMENSION(:)       :: USE_FLAGDEP
LOGICAL,           ALLOCATABLE, DIMENSION(:)       :: USEFLAG
LOGICAL,           ALLOCATABLE, DIMENSION(:)       :: WTCORRELATED
DOUBLE PRECISION,  ALLOCATABLE, DIMENSION(:)       :: WTCOS
```

Explanation of variables:

GROUP contains the name of the group to which each dependent belongs. It is allocated and populated in the private subroutine DEP_INI_FILTERLIST, which is called from DEP_INI_STORE.

GROUPNAM contains the names of groups of dependents.

GROUPNOBS contains the group name for each observation or prediction.

IGPTYPE contains an indicator of the dependent type for each group of dependents. Valid values of IGPTYPE and their meanings include:

0 — dependents are unused,

1 — dependents are used as observations, and

2 — dependents are used as predictions.

LLPTRDEP points to an input-block data structure of dependent data. Each list in the data structure contains information for one dependent. LLPTRDEP is associated in a call to UTL_READBLOCK from DEP_INI_READ.

LLPTRDEPGP points to an input-block data structure of information for dependent groups. Each list in the data structure contains information for one dependent group.

LLPTRDEPGP is associated in a call to UTL_READBLOCK from DEP_INI_READ.

MODEXTVAL contains the simulated equivalent of an observation.

MODDERVAL contains the derived equivalent of an observation.

NONDETVAL contains logical values with the following meanings: True indicates the corresponding value in the OBSVAL array is a detection limit; false indicates the value is a measured value.

NPREDGPS is the number of prediction groups.

NTOTOBS is the total number of observations.

NTOTPRED is the total number of predictions.

NUSEDEP is the number of used dependents.

NUSEOBS is the number of used observations.

NUSEPRED is the number of used predictions.

OBSDEROBSNUM is the order number of derived observations.

OBSEXTNAM contains the names of extracted observations.

OBSEXTOBSNUM is the order number of extracted observation.

OBSVAL is intended to contain observation values. It is allocated and populated in DEP_INI_FILTERLIST, which is called from DEP_INI_STORE.

USE_FLAGDEP contains a temporary flag allocated in DEP_INI_COUNT indicating if the observation is used in the analysis. The alternative is that it is used to calculate a derived observation, which is used in the analysis. It is deallocated after reading and sorting of dependents I complete at the end of DEP_INI_STORE.

USEFLAG contains a flag indicating whether the group is used in the analysis.

WTCORRELATED is intended to indicate for each observation whether or not the observation is correlated with other observations in the same observation group. It is allocated and populated in DEP_INI_FILTERLIST, which is called from DEP_INI_STORE.

WTCOS contains values associated with the keyword WTOSCONSTANT in the input block of dependent data, as defined in Table 16-21.

Private Data

Declaration of variables:

```
CHARACTER(LEN=12),              ALLOCATABLE, DIMENSION(:) :: COVMATNAM
CHARACTER(LEN=MAX_STRING_LEN), ALLOCATABLE, DIMENSION(:) :: DERDEPEQN
LOGICAL,                        ALLOCATABLE, DIMENSION(:) :: EQUATION_FLAG
TYPE (LLIST), POINTER                                     :: LLPTRPRED
TYPE (LLIST), POINTER                                     :: LLPTRPREDGP
INTEGER                                                   :: NDEPGPS
INTEGER                                                   :: NDERDEP
INTEGER                                                   :: NEXTDEP
INTEGER, PARAMETER                                        :: NOBSCOLS = 6
INTEGER                                                   :: NOBSGPS
INTEGER                                                   :: NTOTDEP
INTEGER,                        ALLOCATABLE, DIMENSION(:) :: NUMINGROUP
CHARACTER(LEN=LENDNAM),         ALLOCATABLE, DIMENSION(:) :: OBSALLNAM
CHARACTER(LEN=LENDNAM),         ALLOCATABLE, DIMENSION(:) :: OBSALLNAMLC
CHARACTER(LEN=40),                           DIMENSION(NOBSCOLS,2) :: OBSCOL
CHARACTER(LEN=LENDNAM),         ALLOCATABLE, DIMENSION(:) :: OBSDERNAM
INTEGER,                        ALLOCATABLE, DIMENSION(:) :: PLOTSYMBOL
DOUBLE PRECISION,               ALLOCATABLE, DIMENSION(:) :: VARIANCE
DOUBLE PRECISION,               ALLOCATABLE, DIMENSION(:) :: WTMULTIPLIER
```

Appendix H: Dependents (DEP) Module of Chapter 10

Explanation of variables:

COVMATNAM contains the names of variance-covariance matrices for groups of dependents.

DERDEPEQN contains the derived dependent equations in text format. It is first allocated in DEP_INI_COUNT for all all observations, then deallocated after evaluating each observation to see if it has an equation, and finally reallocated to the number of derived observations in DEP_INI_ALLOC. This stepped approached makes it possible for the user to mix directly extracted and derived observations in the OBSERVATION_DATA table.

EQUATION_FLAG by default is false and is confirmed false in DEP_INI_COUNT if the DERDEPEQN = "_", and otherwise is set to true.

LLPTRPRED points to an input-block data structure of prediction data. Each list in the data structure contains information for one prediction. LLPTRPRED is associated in a call to UTL_READBLOCK from DEP_INI_READ.

LLPTRPREDGP points to an input-block data structure of information for prediction groups. Each list in the data structure contains information for one prediction group. LLPTRPREDGP is associated in a call to UTL_READBLOCK from DEP_INI_READ.

NDEPGPS is the number of dependent groups.

NDERDEP is the number of derived dependents.

NEXTDEP is the number of dependents to be extracted from model-output files.

NOBSCOLS is the default number of columns for the Observation_Data and Prediction_Data input blocks.

NOBSGPS is the number of dependent groups.

NTOTDEP is the total number of dependents.

NUMINGROUP is the number of members in each dependent group.

OBSALLNAM contains the names of all observations.

OBSALLNAMLC contains the names of all observations, converted to lowercase.

OBSCOL contains the default column names for the Observation_Data and Prediction_Data input blocks.

OBSDERNAM contains the names of derived observations.

PLOTSYMBOL contains an identifying integer associated with each dependent group, which is written to data-exchange files. One use for this identifying integer would be to associate plot symbols with groups for a graphics appliction that uses data-exchange file(s) as input.

VARIANCE contains the variance associated with each dependent.

WTMULTIPLIER contains, for each group of dependents, a value that multiplies the weight of each member of the group.

Public Subroutines and Generic Interface – Initialize (INI)

This section documents subroutines and a generic interface that are used primarily for initialization. They are listed alphabetically as follows:

Subroutine DEP_INI_ALLOC
Interface   DEP_INI_NAMCHK
Subroutine DEP_INI_READ
Subroutine DEP_INI_STORE

215

Appendix H: Dependents (DEP) Module of Chapter 10

Subroutine DEP_INI_ALLOC

Description: This subroutine allocates and initializes the module arrays MODEXTVAL, OBSEXTNAM, OBSEXTOBSNUM, DERDEPEQN, MODDERVAL, OBSERNAM, and OBSDEROBSNUM.
Argument list: Empty

Interface DEP_INI_NAMCHK

Description: This generic subroutine interface calls UTL_SHELLSORT to alphabetize the names of the dependents stored in the OBSALLNAM module array or in the NAMES1 array, then checks adjacent names to determine if they are unique, and writes messages regarding nonunique names. The comparison is case-insensitive.
Argument list:
(IOUT, ND1, NAMES1)

Declarations for argument-list variables:
```
INTEGER,                                INTENT(IN) :: IOUT
INTEGER,                                INTENT(IN) :: ND1
CHARACTER(LEN=LENDNAM), DIMENSION(ND1), INTENT(IN) :: NAMES1
```

Explanation of arguments:
IOUT is the unit number associated with the file to which output is to be written.
ND1 is the number of elements in the NAMES1 array.
NAMES1 contains names to be checked.
Note:
ND1 and NAMES1 must either both be present or both be absent. If ND1 and NAMES1 are absent, the OBSALLNAM array is checked. If present, the NAMES1 array is checked.

Subroutine DEP_INI_READ

Description: This subroutine calls UTL_READBLOCK to read up to three input blocks; of these blocks, the first and third blocks are optional and the second block is required. The input blocks to be read are related to observations if the IDEPTYPE argument equals 1, and the blocks to be read are related to predictions if IDEPTYPE equals 2. The first block is optional. If present, it must be labeled "Observation_Groups" if IDEPTYPE equals 1; it is expected to contain information related to observation groups. The first block must be labeled "Prediction_Groups" if IDEPTYPE equals 2; it is expected to contain information related to prediction groups. Regardless of the value of IDEPTYPE, the keyitem for the first input block is "GROUPNAME". The second block is required and must be labeled "Observation_Data" if IDEPTYPE equals 1; it is expected to contain information related to specific observations. The second block must be labeled "Prediction_Data" if IDEPTYPE equals 2; it is expected to contain information related to specific predictions. The keyitem for the second input block is "OBSNAME" if IDEPTYPE equals 1 or "PREDNAME" if IDEPTYPE equals 2. The third block, if present, must be labeled "Derived_Observations" if IDEPTYPE equals 1 or "Derived_Predictions" if IDEPTYPE equals 2. The keyitem for the third block is "OBSNAME" if IDEPTYPE equals 1 or "PREDNAME" if IDEPTYPE equals 2. The data are stored in arrays declared as private in the Dependents Module. The input-block data structure

216

pointed to by LLPTROBS is sent to DEP_INI_FILTERLIST for extraction of values associated with recognized keywords.

Argument list:
```
(IDEPTYPE, INUNIT, IOUT, NCOVMAT, NOBS, NDDEP, NEDEP, NTDEP)
```

Declarations for argument-list variables:
```
INTEGER, INTENT(IN)    :: IDEPTYPE
INTEGER, INTENT(IN)    :: INUNIT
INTEGER, INTENT(IN)    :: IOUT
INTEGER, INTENT(INOUT) :: NCOVMAT
INTEGER, INTENT(OUT)   :: NOBS
INTEGER, INTENT(OUT)   :: NDDEP
INTEGER, INTENT(OUT)   :: NEDEP
INTEGER, INTENT(OUT)   :: NTDEP
```

Explanation of arguments:

    IDEPTYPE is a flag that determines if data to be read by the subroutine are related to observations or predictions. If IDEPTYPE equals 1, the data are related to observations; if IDEPTYPE equals 2, the data are related to predictions.

    INUNIT is the unit number associated with the file from which input is to be read.

    IOUT is the unit number associated with the file to which output is to be written.

    NCOVMAT is the number of variance-covariance matrices to be read.

    NOBS is the number of observations or predictions that are to be used.

    NDDEP is the number of derived observations or predictions.

    NEDEP is the number of extracted observations or predictions.

    NTDEP is the total number of observations or predictions.

Note:

    The contents of the input-block data structure pointed to by LLPTRGPOBS are copied into the lists pointed to by LLPTROBS using subroutine UTL_GROUPLIST, matched according to the keyword GROUPNAME.

Subroutine DEP_INI_STORE

Description: This subroutine allocates and populates arrays for storing observation-related or prediction-related information. The data are stored in arrays declared as private in the Dependents Module.

Argument list:
```
(IDEPTYPE, IOUT, NCOVMAT, NOBS, COVMATARR, NEQNPTR, OBSNAM, DTLA, WTMAT,
WTMATSQR)
```

Declarations for argument-list variables:
```
INTEGER,                                INTENT(IN)    :: IDEPTYPE
INTEGER,                                INTENT(IN)    :: IOUT
INTEGER,                                INTENT(IN)    :: NCOVMAT
INTEGER,                                INTENT(IN)    :: NOBS
TYPE (CDMATRIX), DIMENSION(NCOVMAT),    INTENT(IN)    :: COVMATARR
INTEGER,                                INTENT(IN)    :: NEQNPTR
CHARACTER(LEN=LENDNAM), DIMENSION(NOBS), INTENT(INOUT) :: OBSNAM
DOUBLE PRECISION,                       INTENT(OUT)   :: DTLA
TYPE (CDMATRIX),                        INTENT(OUT)   :: WTMAT
TYPE (CDMATRIX),                        INTENT(OUT)   :: WTMATSQR
```

Explanation of arguments:

IDEPTYPE is a flag that determines if data to be stored by the subroutine are related to observations or predictions. If IDEPTYPE equals 1, the data are related to observations; if IDEPTYPE equals 2, the data are related to predictions.

IOUT is the unit number associated with the file to which output is to be written.

NCOVMAT is the number of variance-covariance matrices stored in the COVMATARR array.

NOBS is the number of observations or predictions.

COVMATARR contains, as a minimum, a variance-covariance matrix for each group of observations or predictions that is listed by name in the GROUP array and is identified as being correlated by TRUE values in the WTCORRELATED array. COVMATARR may contain additional matrices that are not listed in the GROUP array.

NEQNPTR points to the position before the first equation to be used such that NEQNPTR+1 is the first equation.

OBSNAM contains the observation or prediction names.

DTLA is the natural log of the determinant of the inverse of the weight matrix.

WTMAT is the weight matrix.

WTMATSQR is the square root of the weight matrix.

Public Subroutine – Extract Values (EXT)

Subroutine DEP_EXT_DER

Description: This subroutine uses the extracted values to calculate the derived dependents.
Argument list:
```
(IOUT, NDDEP, NEDEP, NEQNPTR, DEREXTVAL, DEPDERVAL)
```

Declarations for argument-list variables:
```
INTEGER,                              INTENT(IN)  :: IOUT
INTEGER,                              INTENT(IN)  :: NDDEP
INTEGER,                              INTENT(IN)  :: NEDEP
INTEGER,                              INTENT(IN)  :: NEQNPTR
DOUBLE PRECISION, DIMENSION(NEDEP),   INTENT(IN)  :: DEREXTVAL
DOUBLE PRECISION, DIMENSION(NDDEP),   INTENT(OUT) :: DEPDERVAL
```

Explanation of arguments:

IOUT is the unit number associated with the file to which output is to be written.

NDDEP is the number of derived dependents.

NEDEP is the number of extracted dependents.

NEQNPTR points to the position before the first equation to be used such that NEQNPTR+1 is the first equation.

DEREXTVAL contains the extracted values of the extracted dependents.

DEPDERVAL contains the calculated values of the derived dependents.

Public Subroutines – Use Extracted Values (UEV)

This section documents subroutines that are used primarily to use the extracted values. They are listed alphabetically as follows:

Subroutine DEP_UEV_DX_READ_OS

Subroutine DEP_UEV_DX_READ_P
Subroutine DEP_UEV_DX_READ_PNUM
Subroutine DEP_UEV_DX_READ_WS
Subroutine DEP_UEV_DX_WRITE_OS
Subroutine DEP_UEV_DX_WRITE_P
Subroutine DEP_UEV_DX_WRITE_R
Subroutine DEP_UEV_DX_WRITE_SS
Subroutine DEP_UEV_DX_WRITE_W
Subroutine DEP_UEV_DX_WRITE_WS
Subroutine DEP_UEV_DX_WRITE_WW
Subroutine DEP_UEV_RESIDUALS
Subroutine DEP_UEV_WRITEOBSTABLE

Subroutine DEP_UEV_DX_READ_OS

Description: This subroutine reads a data-exchange file of type "_os" (Table 17-2).
Argument list:

```
(NTOT, OUTNAM, SIMEQV, OBSVAL, PLOTSYM, OBSNAME)
```

Declarations for argument-list variables:

```
INTEGER,                                INTENT(IN)  :: NTOT
CHARACTER(LEN=MAX_STRING_LEN),          INTENT(IN)  :: OUTNAM
DOUBLE PRECISION, DIMENSION(NTOT),      INTENT(OUT) :: SIMEQV
DOUBLE PRECISION, DIMENSION(NTOT),      INTENT(OUT) :: OBSVAL
INTEGER, DIMENSION(NTOT),               INTENT(OUT) :: PLOTSYM
CHARACTER(LEN=LENDNAM), DIMENSION(NTOT), INTENT(OUT) :: OBSNAME
```

Explanation of arguments:
    NTOT is number records of dependents in the _os file.
    OUTNAM is the root name of the data-exchange file.
    SIMEQV is populated with dependent values read from the _os file.
    OBSVAL is populated with observed or reference values read from the _os file.
    PLOTSYM is populated with plot-symbol values read from the _os file.
    OBSNAME is populated with dependent names read from the _os file.

Subroutine DEP_UEV_DX_READ_P

Description: This subroutine reads a data-exchange file of type "_p" (Table 17-6).
Argument list:

```
(OUTNAM, NPRED, PREDGRP, PREDNAM, PREDVAL, PREDVAR, PLOTSYMBOL)
```

Declarations for argument-list variables:

```
CHARACTER(LEN=MAX_STRING_LEN),          INTENT(IN)  :: OUTNAM
INTEGER,                                INTENT(IN)  :: NPRED
CHARACTER(LEN=12), DIMENSION(NPRED),    INTENT(OUT) :: PREDGRP
CHARACTER(LEN=LENDNAM), DIMENSION(NPRED), INTENT(OUT) :: PREDNAM
DOUBLE PRECISION, DIMENSION(NPRED),     INTENT(OUT) :: PREDVAL
DOUBLE PRECISION, DIMENSION(NPRED),     INTENT(OUT) :: PREDVAR
INTEGER, DIMENSION(NPRED),              INTENT(OUT) :: PLOTSYMBOL
```

Explanation of arguments:

  OUTNAM is the root name of the data-exchange file.

  NPRED is the number of predictions for which data are to be read from the _p file.

  PREDGRP is populated with the names of prediction groups read from the _p file.

  PREDNAM is populated with the names of predictions read from the _p file.

  PREDVAL is populated with the predicted values read from the _p file.

  PREDVAR is populated with the variance values read from the _p file.

  PLOTSYMBOL is populated with the plot symbol values read from the _p file.

Subroutine DEP_UEV_DX_READ_PNUM

Description: This subroutine reads a data-exchange file of type "_p" (Table 17-6) to get number of predictions contained in the file.

Argument list:
```
(IFAIL, OUTNAM, NPRED)
```

Declarations for argument-list variables:
```
INTEGER,                         INTENT(OUT) :: IFAIL
CHARACTER(LEN=MAX_STRING_LEN),   INTENT(IN)  :: OUTNAM
INTEGER,                         INTENT(OUT) :: NPRED
```

Explanation of arguments:

  IFAIL is set to 0 if the file is opened successfully or is set to 1 if it is not.

  OUTNAM is the root name of the data-exchange file.

  NPRED is the number of predictions contained in the _p file.

Subroutine DEP_UEV_DX_READ_WS

Description: This subroutine reads a data-exchange file of type "_ws" (Table 17-2) of simulated equivalents and corresponding weighted residuals.

Argument list:
```
(NTOT, OUTNAM, SIMEQV, WTRES, PLOTSYM, OBSNAME)
```

Declarations for argument-list variables:
```
INTEGER,                                    INTENT(IN)  :: NTOT
CHARACTER(LEN=MAX_STRING_LEN),              INTENT(IN)  :: OUTNAM
DOUBLE PRECISION, DIMENSION(NTOT),          INTENT(OUT) :: SIMEQV
DOUBLE PRECISION, DIMENSION(NTOT),          INTENT(OUT) :: WTRES
INTEGER, DIMENSION(NTOT),                   INTENT(OUT) :: PLOTSYM
CHARACTER(LEN=LENDNAM), DIMENSION(NTOT),    INTENT(OUT) :: OBSNAME
```

Explanation of arguments:

  NTOT is the number of observations for which simulated equivalents are to be read.

  OUTNAM is the root name of the data-exchange file.

  SIMEQV is populated with simulated-equivalent values read from the _ws file.

  WTRES is populated with weighted residuals read from the _ws file.

  PLOTSYM is populated with plot-symbol values read from the _ws file.

  OBSNAME is populated with observation names read from the _ws file.

Appendix H: Dependents (DEP) Module of Chapter 10

Subroutine DEP_UEV_DX_WRITE_OS

Description: This subroutine writes model-simulated and observed values to a data-exchange file of type _os (Table 17-2).

Argument list:
```
(IUOS, NOBS, MODELVAL, OBSNAM, OBSVAL, WRITEHEADER, IPLOTOPT)
```

Declarations for argument-list variables:
```
INTEGER,                                      INTENT(IN) :: IUOS
INTEGER,                                      INTENT(IN) :: NOBS
DOUBLE PRECISION,         DIMENSION(NOBS), INTENT(IN) :: MODELVAL
CHARACTER(LEN=LENDNAM),   DIMENSION(NOBS), INTENT(IN) :: OBSNAM
DOUBLE PRECISION,         DIMENSION(NOBS), INTENT(IN) :: OBSVAL
LOGICAL,                                      INTENT(IN) :: WRITEHEADER
INTEGER, OPTIONAL,        DIMENSION(NOBS), INTENT(IN) :: IPLOTOPT
```

Explanation of arguments:

IUOS is the unit number on which a data-exchange file with the extentsion "_os" has been opened for writing.

NOBS is the number of observations to be written.

MODELVAL contains the simulated-equivalent values.

OBSNAM contains the observation names.

OBSVAL contains the observed values.

WRITEHEADER is a flag to indicate if a header line should be written before the rows of data.

IPLOTOPT contains plot-symbol values associated with the observations.

Notes:

1. A properly constructed _os file has a header line at the top. The first call to DEP_UEV_DX_WRITE_OS should have WRITEHEADER specified as true. Subsequent calls to write additional data to the same file should have WRITEHEADER specified as false.

2. If IPLOTOPT is absent, values from the PLOTSYMBOL module array are written. Use of this option requires that the PLOTSYMBOL array be previously populated by a call to DEP_INI_READ.

Subroutine DEP_UEV_DX_WRITE_P

Description: This subroutine writes predicted values to a data-exchange file of type _p (Table 17-6).

Argument list:
```
(IUP, NPRED, MODELVAL, PREDNAM, WRITEHEADER, IPLOTOPT)
```

Declarations for argument-list variables:
```
INTEGER,                                      INTENT(IN) :: IUP
INTEGER,                                      INTENT(IN) :: NPRED
DOUBLE PRECISION,         DIMENSION(NPRED), INTENT(IN) :: MODELVAL
CHARACTER(LEN=LENDNAM),   DIMENSION(NPRED), INTENT(IN) :: PREDNAM
LOGICAL,                                      INTENT(IN) :: WRITEHEADER
INTEGER, OPTIONAL,        DIMENSION(NPRED), INTENT(IN) :: IPLOTOPT
```

Explanation of arguments:

IUP is the unit number on which a data-exchange file with the extension "_p" has been opened for writing.

NPRED is the number of predictions.

MODELVAL contains the predicted values.

PREDNAM contains the prediction names.

WRITEHEADER is a flag to indicate if a header line should be written before the rows of data.

IPLOTOPT contains plot-symbol values associated with the observations.

Subroutine DEP_UEV_DX_WRITE_R

Description: This subroutine writes model-simulated and observed values to a data-exchange file of type _r (Table 17-2).

Argument list:
```
(IUR, NOBS, MODELVAL, OBSNAM, OBSVAL, WRITEHEADER, IPLOTOPT)
```

Declarations for argument-list variables:
```
INTEGER,                                     INTENT(IN) :: IUR
INTEGER,                                     INTENT(IN) :: NOBS
DOUBLE PRECISION,        DIMENSION(NOBS), INTENT(IN) :: MODELVAL
CHARACTER(LEN=LENDNAM), DIMENSION(NOBS), INTENT(IN) :: OBSNAM
DOUBLE PRECISION,        DIMENSION(NOBS), INTENT(IN) :: OBSVAL
LOGICAL,                                     INTENT(IN) :: WRITEHEADER
INTEGER, OPTIONAL,       DIMENSION(NOBS), INTENT(IN) :: IPLOTOPT
```

Explanation of arguments:

IUR is the unit number on which a data-exchange file with the extension "_r" has been opened for writing.

NOBS is the number of observations.

MODELVAL contains the values of simulated equivalents to the observations.

OBSNAM contains the observation names.

OBSVAL contains the observed values.

WRITEHEADER is a flag to indicate if a header line should be written before the rows of data.

IPLOTOPT contains plot-symbol values associated with the observations.

Subroutine DEP_UEV_DX_WRITE_SS

Description: This subroutine writes an "_ss" data-exchange file (Table 17-5), which contains a header and the sum of squared weighted residuals for observations only, for prior information only, and for observations and prior information combined, for each iteration.

Argument list:
```
(MAXITER, INCLU, ITERP, OUTNAM, RSQA, RSPA)
```

Declarations for argument-list variables:
```
INTEGER,                                       INTENT(IN) :: MAXITER
INTEGER,             DIMENSION(MAXITER+1), INTENT(IN) :: INCLU
INTEGER,                                       INTENT(IN) :: ITERP
CHARACTER(LEN=MAX_STRING_LEN),             INTENT(IN) :: OUTNAM
DOUBLE PRECISION, DIMENSION(MAXITER+1), INTENT(IN) :: RSQA
DOUBLE PRECISION, DIMENSION(MAXITER+1), INTENT(IN) :: RSPA
```

Explanation of arguments:

MAXITER is the maximum number of parameter iterations before the code terminates.

INCLU contains the number of observations used in each iteration.

ITERP is the current iteration number.

OUTNAM is the root file name for output files.

RSQA is the sum of squared weighted residuals for observations only.

RSPA is the sum of squared weighted residuals for observations and prior information combined.

Subroutine DEP_UEV_DX_WRITE_W

Description: This subroutine writes weighted-residual values for observations to a data-exchange file of type _w (Table 17-2).

Argument list:
```
(IUW, NOBS, WTDRESIDS, OBSNAM, WRITEHEADER, IPLOTOPT)
```

Declarations for argument-list variables:
```
INTEGER,                                      INTENT(IN) :: IUW
INTEGER,                                      INTENT(IN) :: NOBS
DOUBLE PRECISION,         DIMENSION(NOBS), INTENT(IN) :: WTDRESIDS
CHARACTER(LEN=LENDNAM), DIMENSION(NOBS), INTENT(IN) :: OBSNAM
LOGICAL,                                      INTENT(IN) :: WRITEHEADER
INTEGER, OPTIONAL,        DIMENSION(NOBS), INTENT(IN) :: IPLOTOPT
```

Explanation of arguments:

IUW is the unit number on which a data-exchange file with the extension "_w" has been opened for writing.

NOBS is the number of observations.

WTDRESIDS contains the weighted residuals.

OBSNAM contains the observation names.

WRITEHEADER is a flag to indicate if a header line should be written before the rows of data.

IPLOTOPT contains plot-symbol values associated with the observations.

Subroutine DEP_UEV_DX_WRITE_WS

Description: This subroutine writes model-simulated and weighted-residual values for observations to a data-exchange file of type _ws (Table 17-2).

Argument list:
```
(IUWS, NOBS, MODELVAL, OBSNAM, WTDRESIDS, WRITEHEADER, IPLOTOPT)
```

Declarations for argument-list variables:
```
INTEGER,                                      INTENT(IN) :: IUWS
INTEGER,                                      INTENT(IN) :: NOBS
DOUBLE PRECISION,         DIMENSION(NOBS), INTENT(IN) :: MODELVAL
CHARACTER(LEN=LENDNAM), DIMENSION(NOBS), INTENT(IN) :: OBSNAM
DOUBLE PRECISION,         DIMENSION(NOBS), INTENT(IN) :: WTDRESIDS
LOGICAL,                                      INTENT(IN) :: WRITEHEADER
INTEGER, OPTIONAL,        DIMENSION(NOBS), INTENT(IN) :: IPLOTOPT
```

Explanation of arguments:

IUWS is the unit number on which a data-exchange file with the extension "_ws" has been opened for writing.

NOBS is the number of observations.

MODELVAL contains the model-simulated values.

OBSNAM contains the observation names.

WTDRESIDS contains the weighted residuals.

WRITEHEADER is a flag to indicate if a header line should be written before the rows of data.

IPLOTOPT contains plot-symbol values associated with the observations.

Subroutine DEP_UEV_DX_WRITE_WW

Description: This subroutine writes weighted model-simulated and weighted observed values for observations to a data-exchange file of type _ww (Table 17-2).

Argument list:

```
(IUWW, NOBS, OBSNAM, WTDOBS, WTDSIM, WRITEHEADER, IPLOTOPT)
```

Declarations for argument-list variables:

```
INTEGER,                                      INTENT(IN) :: IUWW
INTEGER,                                      INTENT(IN) :: NOBS
CHARACTER(LEN=LENDNAM), DIMENSION(NOBS), INTENT(IN) :: OBSNAM
DOUBLE PRECISION,       DIMENSION(NOBS), INTENT(IN) :: WTDOBS
DOUBLE PRECISION,       DIMENSION(NOBS), INTENT(IN) :: WTDSIM
LOGICAL,                                      INTENT(IN) :: WRITEHEADER
INTEGER, OPTIONAL,      DIMENSION(NOBS), INTENT(IN) :: IPLOTOPT
```

Explanation of arguments:

IUWW is the unit number on which a data-exchange file with the extension "_ww" has been opened for writing.

NOBS is the number of observations.

OBSNAM contains the observation names.

WTDOBS contains the weighted observed values.

WTDSIM contains the weighted model-simulated values.

WRITEHEADER is a flag to indicate if a header line should be written before the rows of data.

IPLOTOPT contains plot-symbol values associated with the observations.

Subroutine DEP_UEV_RESIDUALS

Description: This subroutine calculates residuals and weighted residuals and stores the results into 1-D arrays.

Argument list:

```
(NOBS, MODELVAL, WTMATWQR, RESIDS, WTDRESIDS)
```

Declarations for argument-list variables:

```
INTEGER,                                  INTENT(IN)  :: NOBS
DOUBLE PRECISION, DIMENSION(NOBS), INTENT(IN)  :: MODELVAL
TYPE (CDMATRIX),                        INTENT(IN)  :: WTMATSQR
DOUBLE PRECISION, DIMENSION(NOBS), INTENT(OUT) :: RESIDS
DOUBLE PRECISION, DIMENSION(NOBS), INTENT(OUT) :: WTDRESIDS
```

Explanation of arguments:

NOBS is the number of observations.

MODELVAL contains the simulated-equivalent values.

WTMATSQR is the square root of the weight matrix.

RESIDS contains the residuals.

WTDRESIDS contains the weighted residuals.

Notes:

1. Each residual is calculated as (observed value – simulated equivalent).
2. The WTDRESIDS array is calculated as the matrix-multiplication product of the WTMATSQR matrix times the RESIDS array, where the RESIDS array is treated as a column vector.

Subroutine DEP_UEV_WRITEOBSTABLE

Description: This subroutine writes observed and simulated-equivalent values as a table, showing residuals and weighted residuals.

Argument list:

```
(IOUT, NOBS, MODELVAL, OBSNAM, RESIDS, WTDRESIDS)
```

Declarations for argument-list variables:

```
INTEGER,                                     INTENT(IN) :: IOUT
INTEGER,                                     INTENT(IN) :: NOBS
DOUBLE PRECISION, DIMENSION(NOBS),           INTENT(IN) :: MODELVAL
CHARACTER(LEN=LENDNAM), DIMENSION(NOBS),     INTENT(IN) :: OBSNAM
DOUBLE PRECISION, DIMENSION(NOBS),           INTENT(IN) :: RESIDS
DOUBLE PRECISION, DIMENSION(NOBS),           INTENT(IN) :: WTDRESIDS
```

Explanation of arguments:

IOUT is the unit number associated with the file to which output is to be written.

NOBS is the number of observations.

MODELVAL contains the simulated-equivalent values.

OBSNAM contains the observation names.

RESIDS contains the residuals.

WTDRESIDS contains the weighted residuals.

Public Subroutine – Cleanup (CLN)

Subroutine DEP_CLN

Description: This subroutine deallocates all allocatable arrays that are stored in the Dependents Module.

Argument list: Empty

Other Public Subroutines and a Function

This section documents subroutines and a function. They are listed alphabetically as follows:
Subroutine DEP_DX_WRITE_GM
Subroutine DEP_GET_GROUP
Function   DEP_GET_PLOTSYMBOL

Subroutine DEP_DX_WRITE_GM

Description: This subroutine writes dependent names and corresponding group names to a data-exchange file of type _gm (Table 17-2).
Argument list:
```
(IUGM, NOBS, OBSNAM, WRITEHEADER)
```

Declarations for argument-list variables:
```
INTEGER,                                 INTENT(IN) :: IUGM
INTEGER,                                 INTENT(IN) :: NOBS
CHARACTER(LEN=LENDNAM), DIMENSION(NOBS), INTENT(IN) :: OBSNAM
LOGICAL,                                 INTENT(IN) :: WRITEHEADER
```

Explanation of arguments:
IUGM is the unit number on which a data-exchange file with the extension "_gm" has been opened for writing.
NOBS is the number of dependents.
OBSNAM contains the names of the dependents.
WRITEHEADER is a flag to indicate if a header line should be written before the rows of data.

Subroutine DEP_GET_GROUP

Description: This function returns the group name and plot symbol associated with an observation number.
Argument list:
```
(N, GPNAM, IPLOT, DEPSET)
```

Declarations for argument-list variables:
```
INTEGER,                      INTENT(IN)  :: N
CHARACTER(LEN=12),            INTENT(OUT) :: GPNAM
INTEGER,                      INTENT(OUT) :: IPLOT
CHARACTER(LEN=*), OPTIONAL,   INTENT(IN)  :: DEPSET
```

Explanation of arguments:
N is the observation number. See note.
GPNAM is the group name.
IPLOT is the plot symbol.
DEPSET, if present, is a string for which "ALL" and "USED" are the only valid values. See note.

Note:

If DEPSET is "ALL," then N is interpreted as being the observation number out of all NTOTDEP dependents. If DEPSET is "USED," then N is interpreteted as being the observation number out of NUSEDEP used dependents.

Function DEP_GET_PLOTSYMBOL

Description: This function returns the plot-symbol value stored in the private array PLOTSYMBOL for group GNAM if found in array GROUPNAM.

Argument list:
```
(GNAM, IDEFAULT)
```

Result:
```
IPLOT
```

Declarations for argument-list and result variables:
```
CHARACTER(LEN=12), INTENT(IN) :: GNAM
INTEGER,           INTENT(IN) :: IDEFAULT
INTEGER                       :: IPLOT
```

Explanation of arguments and result:

GNAM is the name of the group for which the plot-symbol value is needed.

IDEFAULT is the default plot-symbol value to be returned if GNAM is not in the list of groups

IPLOT is the plot-symbol value associated with group GNAM if GNAM is in GROUPNAM, which is private to the Dependents Module; if GNAM is not in GROUPNAM, IPLOT is set to IDEFAULT.

Appendix I: Prior-Information (PRI) Module of Chapter 11

Public Data

Declaration of variables:
```
TYPE (LLIST), POINTER                                        :: LLPTRGPPRIOR
TYPE (LLIST), POINTER                                        :: LLPTRPRIOR
DOUBLE PRECISION,        ALLOCATABLE, DIMENSION(:)   :: MODELPRIVAL
INTEGER,                 ALLOCATABLE, DIMENSION(:)   :: PLOTSYMBOLPRI
DOUBLE PRECISION,        ALLOCATABLE, DIMENSION(:,:) :: PRIEQN
INTEGER,                 ALLOCATABLE, DIMENSION(:)   :: PRILN
CHARACTER(LEN=LENDNAM), ALLOCATABLE, DIMENSION(:)   :: PRINAM
CHARACTER(LEN=LENDNAM), ALLOCATABLE, DIMENSION(:)   :: PRINAMLC
DOUBLE PRECISION,        ALLOCATABLE, DIMENSION(:)   :: PRIVAL
LOGICAL,                 ALLOCATABLE, DIMENSION(:)   :: PRIWTCORR
TYPE (CDMATRIX)                                             :: PRIWTMAT
TYPE (CDMATRIX)                                             :: PRIWTMATSQR
DOUBLE PRECISION,        ALLOCATABLE, DIMENSION(:)   :: RESIDSPRI
DOUBLE PRECISION,        ALLOCATABLE, DIMENSION(:)   :: WTDRESIDSPRI
```

Explanation of variables:
LLPTRGPPRIOR points to an input-block data structure of information for prior groups. Each list contains information for one prior group.
LLPTRPRIOR points to an input-block data structure of prior-information data. Each list contains information for one prior-information equation.
MODELPRIVAL contains the values of the prior-information equation, evaluated for the current parameters.
PLOTSYMBOLPRI contains the plot symbol for the prior-information equation.
PRIEQN contains sensitivities for prior-information equations with respect to adjustable parameters.
PRILN contains a flag indicating whether the prior information is on a transformed parameter (0 if not transformed, 1 if transformed).
PRINAM is intended to contain the name of each prior-information equation. It is allocated and populated in PRI_INI_READ. The equation names are those associated with the keyword "PRIORNAME" in the Linear_Prior_Information input block.
PRINAMLC contains the prior names in lowercase.
PRIVAL contains the prior-information values.
PRIWTCORR contains a flag indicating whether weights are correlated for the group.
PRIWTMAT contains the weight matrix for all prior-information equations.
PRIWTMATSQR contains the square root weight matrix for all prior-information equations.
RESIDSPRI contains the residuals for prior information.
WTDRESIDSPRI contains the weighted residuals for prior information.

Private Data

Declaration of variables:
```
INTEGER, PARAMETER                                          :: NPRICOLS = 6
INTEGER                                                     :: NPRIGPS
INTEGER,                     ALLOCATABLE, DIMENSION(:) :: NUMINPRIGROUP
INTEGER,                     ALLOCATABLE, DIMENSION(:) :: PLOTSYMBOLPRIGRP
CHARACTER(LEN=40),                       DIMENSION(NPRICOLS) :: PRICOL
```

```
CHARACTER(LEN=12),                ALLOCATABLE, DIMENSION(:) :: PRICOVMATNAM
CHARACTER(LEN=MAX_STRING_LEN),    ALLOCATABLE, DIMENSION(:) :: PRIEQTEXT
CHARACTER(LEN=12),                ALLOCATABLE, DIMENSION(:) :: PRIGROUP
CHARACTER(LEN=12),                ALLOCATABLE, DIMENSION(:) :: PRIGROUPNAM
TYPE (CHAR_1D_30),                ALLOCATABLE, DIMENSION(:) :: PRIMEMBER
LOGICAL,                          ALLOCATABLE, DIMENSION(:) :: PRIUSEFLAG
DOUBLE PRECISION,                 ALLOCATABLE, DIMENSION(:) :: PRIVARIANCE
DOUBLE PRECISION,                 ALLOCATABLE, DIMENSION(:) :: PRIWTMULTIPLIER
DOUBLE PRECISION,                 ALLOCATABLE, DIMENSION(:) :: WTDPRI
```

Explanation of variables:
>   NPRICOLS is the number of default columns in the Linear_Prior_Information input block.
>   NPRIGPS is the number of groups of prior information.
>   NUMINPRIGROUP is the number of prior items in each group.
>   PLOTSYMBOLPRIGRP contains the plot symbol for each prior-information group.
>   PRICOL contains the default column names for the Linear_Prior_Information input block.
>   PRICOVMATNAM is the name of the covariance matrix for the prior group.
>   PRIEQTEXT is the text equation for the prior item.
>   PRIGROUP contains the group name for each prior-information equation.
>   PRIGROUPNAM contains the names of the prior groups.
>   PRIMEMBER contains names of prior-information equations that belong to each group.
>   PRIUSEFLAG is a flag indicating whether the prior items will be included in the analysis.
>   PRIVARIANCE is the variance on each prior item.
>   PRIWTMULTIPLIER is a multiplier for all weights in the group.
>   WTDPRI contains the weighted prior-information values.

Public Subroutines – Initialize (INI)

This section documents subroutines that are used primarily for initialization. They are listed alphabetically as follows:
>   Subroutine PRI_INI_ALLOC
>   Subroutine PRI_INI_INSTALL
>   Subroutine PRI_INI_POPX
>   Subroutine PRI_INI_PROCESS
>   Subroutine PRI_INI_READ
>   Subroutine PRI_INI_STORE

Subroutine PRI_INI_ALLOC

Description: This subroutine allocates arrays for prior information.
Argument list:
```
(MPR)
```

Declarations for argument-list variables:
```
INTEGER, INTENT(IN) :: MPR
```

Explanation of arguments:
>   MPR is the number of prior-information equations.

Appendix I: Prior-Information (PRI) Module of Chapter 11

Subroutine PRI_INI_INSTALL

Description: This subroutine installs the prior equations.
Argument list:
```
(IOUT, MPR, NEQNPTR)
```

Declarations for argument-list variables:
```
INTEGER, INTENT(IN) :: IOUT
INTEGER, INTENT(IN) :: MPR
INTEGER, INTENT(IN) :: NEQNPTR
```

Explanation of arguments:
> IOUT is the unit number associated with the file to which output is to be written.
> MPR is the number of prior-information equations.
> NEQNPTR points to the position before the first equation to be used such that NEQNPTR+1 is the first equation.

Subroutine PRI_INI_POPX

Description: This subroutine populates the sensitivity matrix for prior-information equations. If IVERB of the Global Data Module is greater than 3, the prior-information sensitivity matrix and the prior-information weight matrix are written to unit IOUT.
Argument list:
```
(IOUT, IPRC, MPR, NPE, NPT, IPTR, LN, PARNAM, XLOG, XPRI, XPRIFILLED)
```

Declarations for argument-list variables:
```
INTEGER,                                 INTENT(IN)    :: IOUT
INTEGER,                                 INTENT(IN)    :: IPRC
INTEGER,                                 INTENT(IN)    :: MPR
INTEGER,                                 INTENT(IN)    :: NPE
INTEGER,                                 INTENT(IN)    :: NPT
INTEGER, DIMENSION(NPE),                 INTENT(IN)    :: IPTR
INTEGER, DIMENSION(NPT),                 INTENT(IN)    :: LN
CHARACTER(LEN=12), DIMENSION(NPT),       INTENT(IN)    :: PARNAM
DOUBLE PRECISION,                        INTENT(IN)    :: XLOG
DOUBLE PRECISION, DIMENSION(NPE,MPR),    INTENT(INOUT) :: XPRI
LOGICAL,                                 INTENT(INOUT) :: XPRIFILLED
```

Explanation of arguments:
> IOUT is the unit number associated with the file to which output is to be written.
> IPRC is a code that defines the format to be used when the prior-information sensitivity matrix is written. Valid values of IPRC and corresponding output formats are documented in the description of UTL_WRITEMATRIX in the "Data Output" section of Appendix D.
> MPR is the number of prior-information equations.
> NPE is the number of estimated parameters.
> NPT is the total number of parameters.
> IPTR is the position of parameter in the full set of parameters.
> LN is a flag indicating whether the parameter is log-transformed.
> PARNAM contains the parameter names.
> XLOG sensitivity in log space for transformed parameters.
> XPRI sensitivity array for prior information.

230

XPRIFILLED a flag indicating whether the linear prior-information sensitivity array has been filled as it need only be done once.

Subroutine PRI_INI_PROCESS

Description: This subroutine stores and optionally echoes prior-information-related information. It builds a weight matrix for all prior-information equations and stores it in the module structure PRIWTMAT, and it calculates weighted prior estimates and stores it in the module array WTDPRI.

Argument list:
```
(IOUT, IWRITE, MPR, NCOVMAT, COVMATARR, DTLA)
```

Declarations for argument-list variables:
```
INTEGER,                                INTENT(IN)  :: IOUT
LOGICAL,                                INTENT(IN)  :: IWRITE
INTEGER,                                INTENT(IN)  :: MPR
INTEGER,                                INTENT(IN)  :: NCOVMAT
TYPE (CDMATRIX), DIMENSION(NCOVMAT),    INTENT(IN)  :: COVMATARR
DOUBLE PRECISION,                       INTENT(OUT) :: DTLA
```

Explanation of arguments:

IOUT is the unit number associated with the file to which output is to be written.

IWRITE is a flag indicating whether to echo the prior information.

MPR is the number of prior-information equations.

NCOVMAT is the number of variance-covariance matrices stored in the COVMATARR array.

COVMATARR contains, as a minimum, a variance-covariance matrix for each group of prior-information equations that is listed by name in the PRIGROUP array and is identified as being correlated by TRUE values in the PRIWTCORR array. COVMATARR may contain additional matrices that are not listed in the PRIGROUP array.

DTLA is the log-determinant of the matrix.

Subroutine PRI_INI_READ

Description: This subroutine calls UTL_READBLOCK to read two input blocks. The first block is labeled "Prior_Information_Groups"; the REQUIRED argument of UTL_READBLOCK is false, so the "Prior_Information_Groups" input block is optional. If present, the "Prior_Information_Groups" input block is expected to contain information related to groups of prior-information equations. The keyitem for this input block is "GROUPNAME". The second block is the "Linear_Prior_Information" block and also is optional. The keyitem for this input block is "PRIORNAME". If present, it is expected to contain information needed to define individual prior-information equations. If both input blocks are present, the keywords and corresponding values in the "Prior_Information_Groups" block are inserted into the input-block data structure holding data from the "Linear_Prior_Information" block according to matching GROUPNAME values.

Argument list:
```
(INUNIT, IOUT, IWRITE, NPE, NCOVMAT, MPR)
```

Declarations for argument-list variables:
```
INTEGER, INTENT(IN)    :: INUNIT
INTEGER, INTENT(IN)    :: IOUT
```

```
LOGICAL, INTENT(IN)    :: IWRITE
INTEGER, INTENT(IN)    :: NPE
INTEGER, INTENT(INOUT) :: NCOVMAT
INTEGER, INTENT(OUT)   :: MPR
```

Explanation of arguments:

INUNIT is the unit number associated with the file from which input is to be read.

IOUT is the unit number associated with the file to which output is to be written.

IWRITE is a flag indicating whether the number of prior-information equations should be written to IOUT.

NPE is the number of adjustable parameters.

NCOVMAT is the number of variance-covariance matrices to be read.

MPR is the number of prior-information equations read from the "Linear_Prior_Information" input block.

Subroutine PRI_INI_STORE

Description: This routine stores the prior data in arrays.
Argument list:
```
(IOUT, MPR, NPE, NPT, ITRANS, PARNAMLC, PVAL, XPRIFILLED)
```

Declarations for argument-list variables:
```
INTEGER,                            INTENT(IN)    :: IOUT
INTEGER,                            INTENT(IN)    :: MPR
INTEGER,                            INTENT(IN)    :: NPE
INTEGER,                            INTENT(IN)    :: NPT
INTEGER, DIMENSION(NPT),            INTENT(IN)    :: ITRANS
CHARACTER(LEN=12), DIMENSION(NPT),  INTENT(IN)    :: PARNAMLC
DOUBLE PRECISION,  DIMENSION(NPT),  INTENT(IN)    :: PVAL
LOGICAL,                            INTENT(INOUT) :: XPRIFILLED
```

Explanation of arguments:

IOUT is the unit number associated with the file to which output is to be written.

MPR is the number of prior items.

NPE is the number of adjustable parameters.

NPT is the number of parameters.

ITRANS contains the transformation flag for each parameter, as required for subroutine EQN_LINEAR_COEFFS of the Equation Module (Chapter 9).

PARNAMLC contains the parameter names, in all lowercase.

PVAL contains the current parameter values.

XPRIFILLED is a flag indicating whether the linear prior-information sensitivity array has been filled as it need only be done once.

Public Subroutine – Define This Iteration (DEF)

Subroutine PRI_DEF

Description: This subroutine assigns NUMPPL equal to 0.
Argument list:
```
(NUMPPL)
```

Appendix I: Prior-Information (PRI) Module of Chapter 11

Declarations for argument-list variables:
```
INTEGER, INTENT(OUT) :: NUMPPL
```

Explanation of arguments:

NUMPPL is the number of iterations of the PPL loop required to complete the current job.

Public Subroutines – Use Extracted Values (UEV)

This section documents subroutines that are used primarily to use extracted values. They are listed alphabetically as follows:

Subroutine PRI_UEV_ DX_READ_PR
Subroutine PRI_UEV_ DX_WRITE_PR
Subroutine PRI_UEV_ RESIDUALS

Subroutine PRI_UEV_ DX_READ_PR

Description: This subroutine reads prior-information equations from a data-exchange file of type _pr (Table 17-2).
Argument list:
```
(MPR, OUTNAM, PLOTSYM, PRIEQNTXT, PRINAME, PRIVAL)
```

Declarations for argument-list variables:
```
INTEGER,                                     INTENT(IN)  :: MPR
CHARACTER(LEN=MAX_STRING_LEN),               INTENT(IN)  :: OUTNAM
INTEGER, DIMENSION(MPR),                      INTENT(OUT) :: PLOTSYM
CHARACTER(LEN=MAX_STRING_LEN), DIMENSION(MPR), INTENT(OUT) :: PRIEQNTXT
CHARACTER(LEN=LENDNAM), DIMENSION(MPR),        INTENT(OUT) :: PRINAME
DOUBLE PRECISION, DIMENSION(MPR),             INTENT(OUT) :: PRIVAL
```

Explanation of arguments:

MPR is the number of prior-information equations.

OUTNAM is the root name for the data-exchange file.

PLOTSYM is populated with the plot-symbol values.

PRIEQNTXT is populated with the prior-information equations.

PRINAM is populated with the prior-information equation names.

PRIVAL is populated with the prior-information equation values.

Subroutine PRI_UEV_ DX_WRITE_PR

Description: This subroutine writes prior-information equations to a data-exchange file of type _pr (Table 17-2).
Argument list:
```
(MPR, OUTNAM)
```

Declarations for argument-list variables:
```
INTEGER,                     INTENT(IN) :: MPR
CHARACTER(LEN=*),            INTENT(IN) :: OUTNAM
```

Explanation of arguments:
   MPR is the number of prior-information equations.
   OUTNAM is the root name for the data-exchange file.

Subroutine PRI_UEV_ RESIDUALS

Description: This subroutine calculates the residuals on prior information.
Argument list:
```
(IOUT, MPR, NEQNPTR, NPT, PARNAMLC, PVAL)
```

Declarations for argument-list variables:
```
INTEGER,                              INTENT(IN) :: IOUT
INTEGER,                              INTENT(IN) :: MPR
INTEGER,                              INTENT(IN) :: NEQNPTR
INTEGER,                              INTENT(IN) :: NPT
CHARACTER(LEN=12), DIMENSION(NPT),    INTENT(IN) :: PARNAMLC
DOUBLE PRECISION, DIMENSION(NPT),     INTENT(IN) :: PVAL
```

Explanation of arguments:
   IOUT is the unit number associated with the file to which output is to be written.
   MPR is the number of prior-information equations.
   NEQNPTR points to the position before the first equation to be used such that NEQNPTR+1 is
      the first equation
   NPT is the total number of parameters, including adjustable and non-adjustable, primary or
      derived parameters
   PNAMLC contains the parameter names in lower case.
   PVAL contains the parameters to which the PRM coefficients apply.

Public Subroutine – Cleanup (CLN)

Subroutine PRI_CLN

Description: This subroutine deallocates all allocatable arrays that are stored in the Prior-
   Information Module.
Argument list: Empty

Appendix J: Parallel-Processing (PLL) Module of Chapter 12

Public Data

Declaration of variables:

```
LOGICAL                         :: AUTOSTOPRUNNERS = .TRUE.
CHARACTER(LEN=MAX_STRING_LEN)   :: COMMAND = ' '
CHARACTER(LEN=11), PARAMETER    :: FNDISFIN = 'jdispatch.fin'
CHARACTER(LEN=11), PARAMETER    :: FNDISPAR = 'jdispar.rdy'
CHARACTER(LEN=11), PARAMETER    :: FNDISRDY = 'jdispatch.rdy'
CHARACTER(LEN=11), PARAMETER    :: FNRUNDEP = 'jrundep.rdy'
CHARACTER(LEN=12), PARAMETER    :: FNRUNFAIL = 'jrunfail.fin'
CHARACTER(LEN=10), PARAMETER    :: FNRUNFIN = 'jrunner.fin'
CHARACTER(LEN=10), PARAMETER    :: FNRUNRDY = 'jrunner.rdy'
TYPE (LLIST),      POINTER      :: LLPTRPLLCTRL
TYPE (LLIST),      POINTER      :: LLPTRPLLRUN
CHARACTER(LEN=20)               :: SNAME = ' '
DOUBLE PRECISION                :: WAIT = 0.001D0
```

Explanation of variables:

AUTOSTOPRUNNERS is a flag used by subroutine PLL_STOP_RUNNERS which, when true, indicates that the jdispatch.fin file is to instruct active runners to stop execution. If AUTOSTOPRUNNERS is false when PLL_STOP_RUNNERS is called, the jdispatch.fin file contains an instruction that indicates the active runners should reset in preparation for additional instructions in the form of signal files written by the dispatcher program.

COMMAND is the command to be issued by the operating system to initiate a model run.

FNDISFIN contains "jdispatch.fin", which is the name of a signal file that communicates to the runner programs that the dispatcher program is stopping execution.

FNDISPAR contains "jdispar.rdy", which is the name of a signal file that communicates data from the dispatcher program to the runner programs, including values of parameters.

FNDISRDY contains "jdispatch.rdy", which is the name of a signal file that communicates initialization data from the dispatcher program to the runner programs.

FNRUNDEP contains "jrundep.rdy", which is the name of a signal file that communicates values extracted from model-output files from a runner program to the dispatcher program.

FNRUNFAIL contains "jrunfail.fin", which is the name of a signal file that communicates error status and an error message from a runner program to the dispatcher program.

FNRUNFIN contains "jrunner.fin", which is the name of a signal file that communicates to the dispatcher program that a runner program is stopping execution.

FNRUNRDY contains "jrunner.rdy", which is the name of a signal file written by a runner program to communicate to the dispatcher program that the runner is active.

LLPTRPLLCTRL is a pointer to an input-block structure for the "Parallel_Control" input block.

LLPTRPLLRUN is a pointer to an input-block structure for the "Parallel_Runners" input block.

SNAME is the name assigned by the dispatcher program to a one of a set of runners. This variable is assigned and used by a runner program.

WAIT is the time-delay increment (seconds) used as needed in checking for presence of, and reading from, signal files.

Private Data

Declaration of variables:

```
INTEGER,                    ALLOCATABLE, DIMENSION(:,:) :: BTIME
LOGICAL                                                 :: DOPLL
CHARACTER(LEN=MAX_STRING_LEN)                           :: ERRRUNNER
INTEGER,                    ALLOCATABLE, DIMENSION(:)   :: IRUNNERSTAT
INTEGER                                                 :: IVERBRUNNER = 3
INTEGER, PARAMETER                                      :: NRUNNERCOLS = 3
INTEGER                                                 :: NUMRUNNERS
CHARACTER(LEN=20)                                       :: OS_DIS
CHARACTER(LEN=20)                                       :: OS_PLL
CHARACTER(LEN=20)                                       :: OS_TEMP
CHARACTER(LEN=6)                                        :: RENAME_DIS
CHARACTER(LEN=40),                  DIMENSION(NRUNNERCOLS) :: RUNNERCOLS
CHARACTER(LEN=MAX_STRING_LEN), ALLOCATABLE, DIMENSION(:) :: RUNNERDIR
CHARACTER(LEN=20),          ALLOCATABLE, DIMENSION(:)   :: RUNNERNAME
CHARACTER(LEN=MAX_STRING_LEN)                           :: RUNRECFIL
DOUBLE PRECISION,           ALLOCATABLE, DIMENSION(:)   :: RUNTIME
CHARACTER(LEN=6)                                        :: SRENAME = 'ren'
DOUBLE PRECISION                                        :: TIMEOUTFAC
```

Explanation of variables:

BTIME contains data, for each active runner, returned by the DATE_AND_TIME intrinsic subroutine at the time when the "jdispar.rdy" file is written by the dispatcher program.

DOPLL is false by default and is set to true if the PARALLEL keyword of the "Parallel_Control" input block is present and is true.

ERRRUNNER is assigned an error message by a runner, when appropriate. When PLL_RUNNER_STOP is invoked, if ERRRUNNER contains a message, it is written to the "jrunfail.fin" signal file.

IRUNNERSTAT contains a flag indicating the status of each runner. Valid values and meanings are:

Less than 0 – Runner nonresponsive

0 – Initial status

1 – Runner recognized as active

2 – "jdispatch.rdy" file copied to runner directory

3 – "jdispar.rdy" file copied to runner directory; presumably, model is running

4 – "jrundep.rdy" file present in runner directory; model run done

8 – "jdispatch.fin" file copied to runner directory; signal to stop JRUNNER executable

9 – "jrunner.fin" file found, indicating JRUNNER executable has stopped execution

10 – "jrunfail.fin" file found, indicating JRUNNER executable has stopped execution due to error

IVERBRUNNER is assigned the value associated with keyword VERBOSERUNNER of the Parallel_Runners input block. It is written to the "jdispatch.rdy" file, which comminicates the value to the runner program. Its meaning to the runner program is the same as IVERB of the Global Data Module (Appendix C). IVERBRUNNER is initialized to 3, its default value.

NRUNNERCOLS is the default number of columns expected in the "Parallel_Runners" input block.

NUMRUNNERS is the number of runners listed in the "Parallel_Runners" input block.

OS_DIS is the operating system under which the dispatcher program runs (unused in initial release of JUPITER API).

OS_PLL is the opterating system under which bothe the dispatcher and runner programs run.

OS_TEMP temporarily stores an operating-system name.

RENAME_DIS is the command on the operating system under which the dispatcher program runs that is required to rename a file.

RUNNERCOLS contains the default column names for the "Parallel_Runners" input block.

RUNNERDIR contains the name (a relative or absolute pathname) of each directory where the JRUNNER program is (or may be) running.

RUNNERNAME contains a user-specified name for each runner.

RUNRECFIL is the name of a file to which information about parallel processing is to be written (unused in initial release of JUPITER API).

RUNTIME is the expected time (seconds) required for a model run on each runner.

SRENAME is the command that is required to rename a file on the operating system under which a runner program runs. SRENAME is initialized to "ren," its default value.

Public Subroutines – Initialize (INI)

This section documents subroutines that are used primarily for initialization. They are listed alphabetically as follows:

Subroutine PLL_INI_DISPATCHER
Subroutine PLL_INI_RUNNER_DIM
Subroutine PLL_INI_RUNNER_POP

Subroutine PLL_INI_DISPATCHER

Description: This subroutine initializes the Parallel-Processing Module, reads the Parallel_Control and Parallel_Runners input blocks, and allocates and populates arrays with data from the Parallel_Runners input block. It also writes data to the "jdispatch.rdy" file(s), including data needed by a runner program to initialize the Model Input-Output Module. This subroutine is intended to be called from a dispatcher program.

Argument list:
```
(INUNIT, IOUT, LCIS, NCATS, NINSTRUCT, NMIFILE, NMOFILE, NOPNT, NPT, NUMLADV,
NVEXT, PRECISPRO, CATGOR, CINSTSET, EXTNAM, INSTRUCTFILE, LOCINS, MCVUSE,
MIFILE, MOFILE, MRKDL, NW, PARNAM, TFILE, PARACTIVE)
```

Declarations for argument-list variables:
```
INTEGER,                                    INTENT(IN)  :: INUNIT
INTEGER,                                    INTENT(IN)  :: IOUT
INTEGER,                                    INTENT(IN)  :: LCIS
INTEGER,                                    INTENT(IN)  :: NCATS
INTEGER,                                    INTENT(IN)  :: NINSTRUCT
INTEGER,                                    INTENT(IN)  :: NMIFILE
INTEGER,                                    INTENT(IN)  :: NMOFILE
INTEGER,                                    INTENT(IN)  :: NOPNT
INTEGER,                                    INTENT(IN)  :: NPT
INTEGER,                                    INTENT(IN)  :: NUMLADV
INTEGER,                                    INTENT(IN)  :: NVEXT
INTEGER,                                    INTENT(IN)  :: PRECISPRO
CHARACTER(LEN=6),    DIMENSION(NMOFILE),    INTENT(IN)  :: CATGOR
CHARACTER(LEN=1),    DIMENSION(LCIS),       INTENT(IN)  :: CINSTSET
```

237

```
CHARACTER(LEN=LENDNAM), DIMENSION(NVEXT),   INTENT(IN)  :: EXTNAM
CHARACTER (LEN=MAX_STRING_LEN), DIMENSION(NMOFILE), INTENT(IN)  :: INSTRUCTFILE
INTEGER,               DIMENSION(NINSTRUCT), INTENT(IN)  :: LOCINS
LOGICAL,               DIMENSION(NCATS),   INTENT(IN)  :: MCVUSE
CHARACTER(LEN=MAX_STRING_LEN),  DIMENSION(NMIFILE), INTENT(IN)  :: MIFILE
CHARACTER (LEN=MAX_STRING_LEN), DIMENSION(NMOFILE), INTENT(IN)  :: MOFILE
CHARACTER (LEN=1),        DIMENSION(NMOFILE), INTENT(IN)  :: MRKDL
INTEGER,               DIMENSION(NPT),     INTENT(IN)  :: NW
CHARACTER(LEN=12),        DIMENSION(NPT),     INTENT(IN)  :: PARNAM
CHARACTER(LEN=MAX_STRING_LEN),  DIMENSION(NMIFILE), INTENT(IN)  :: TFILE
LOGICAL,                         INTENT(OUT) :: PARACTIVE
```

Explanation of arguments:

INUNIT is the unit number from which input is to be read.

IOUT is the unit number associated with the file to which output is to be written.

LCIS is equal to LENCIS of the Model Input-Output Module.

NCATS is the number of model-calculated value categories.

NINSTRUCT is equal to NINSTR of the Model Input-Output Module.

NMIFILE is equal to NUMINFILE of the Model Input-Output Module.

NMOFILE is equal to NUMOUTFILE of the Model Input-Output Module.

NOPNT is the decimal point protocol setting of the Model Input-Output Module.

NPT is the number of parameters to be substituted into model-input file(s).

NUMLADV is equal to NUML of the Model Input-Output Module.

NVEXT is the number of model-calculated values to be extracted from model-output file(s).

PRECISPRO is equal to PRECPR of the Model Input-Output Module. Its value is either 0 for single precision or not 0 for double precision.

CATGOR contains the model-calculated-value category for each model-output file.

CINSTSET is equal to CISET of the Model Input-Output Module.

EXTNAM contains the names of values to be extracted from model-output file(s).

INSTRUCTFILE contains the names of instruction files stored in the INSFILE array of the Model Input-Output Module.

LOCINS is equal to the LCINS array of the Model Input-Output Module.

MCVUSE indicates, for each model-calculated-value category supported by the application, whether extractions should be performed on model-output files associated with the category.

MIFILE is equal to MODINFILE of the Model Input-Output Module.

MOFILE is equal to MODOUTFILE of the Model Input-Output Module.

MRKDL is equal to MRKDEL of the Model Input-Output Module.

NW contains the number of characters available for writing each parameter to model-input file(s).

PARNAM contains the parameter names.

TFILE contains the names of template file(s); it is equal to the TEMPFILE array of the Model Input-Output Module.

PARACTIVE is assigned as true if the Parallel_Control input block includes the keyword PARALLEL and its value evaluates as true. If PARACTIVE is true, the parallel-processing capability of the Parallel-Processing Module is activated; if false, parallel processing is deactivated.

Appendix J: Parallel-Processing (PLL) Module of Chapter 12

Subroutine PLL_INI_RUNNER_DIM

Description: This subroutine looks for a file named "jdispatch.rdy". When this file is found, the contents of the first part of the file are read, and scalar variables are assigned accordingly. This subroutine is intended to be called from a runner program.

Argument list:
```
(IOUNIT, RESET, LCIS, NCATS, NINSTRUCT, NMIFILE, NMOFILE, NOPNT, NPT, NUMLADV,
NVEXT, PRECISPRO, INIMESS)
```

Declarations for argument-list variables:
```
INTEGER,                        INTENT(IN)    :: IOUNIT
LOGICAL,                        INTENT(INOUT) :: RESET
INTEGER,                        INTENT(OUT)   :: LCIS
INTEGER,                        INTENT(OUT)   :: NCATS
INTEGER,                        INTENT(OUT)   :: NINSTRUCT
INTEGER,                        INTENT(OUT)   :: NMIFILE
INTEGER,                        INTENT(OUT)   :: NMOFILE
INTEGER,                        INTENT(OUT)   :: NOPNT
INTEGER,                        INTENT(OUT)   :: NPT
INTEGER,                        INTENT(OUT)   :: NUMLADV
INTEGER,                        INTENT(OUT)   :: NVEXT
INTEGER,                        INTENT(OUT)   :: PRECISPRO
CHARACTER(LEN=*), OPTIONAL,     INTENT(IN)    :: INIMESS
```

Explanation of arguments:
> IOUNIT is an unused unit number available for use by PLL_INI_RUNNER_DIM. See note below.
>
> RESET is set to true if the "jdispatch.fin" signal file contains the word "RESET", indicating that the runner should reset itself in preparation for another execution of the dispatcher program.
>
> LCIS is equal to LENCIS of the Model Input-Output Module.
>
> NCATS is the number of model-calculated value categories.
>
> NINSTRUCT is equal to NINSTR of the Model Input-Output Module.
>
> NMIFILE is equal to NUMINFILE of the Model Input-Output Module.
>
> NMOFILE is equal to NUMOUTFILE of the Model Input-Output Module.
>
> NOPNT is the decimal point protocol setting of the Model Input-Output Module.
>
> NPT the number of parameters to be substituted into model-input file(s).
>
> NUMLADV is equal to NUML of the Model Input-Output Module.
>
> NVEXT is the number of model-calculated values to be extracted from model-output file(s).
>
> PRECISPRO is equal to PRECPR of the Model Input-Output Module. Its value is either 0 for single precision or not 0 for double precision.
>
> INIMESS is an optional message which, if present, is written to the screen.

Note:
> If the "jdispatch.rdy" signal file is found, PLL_INI_RUNNER_DIM opens the file on unit IOUNIT, reads some of the data in it, and returns, leaving this file open. A subsequent call to PLL_INI_RUNNER_POP should use the same unit number for the IOUNIT argument of PLL_INI_RUNNER_POP so that the remainder of the data in "jdispatch.rdy" can be read by PLL_INI_RUNNER_POP.

Appendix J: Parallel-Processing (PLL) Module of Chapter 12

Subroutine PLL_INI_RUNNER_POP

Description: This subroutine reads the second part of a jdispatch.rdy file and uses the contents to
  populate arrays. It is intended to be called from a runner program.

Argument list:

```
(IOUNIT, LCIS, NCATS, NINSTRUCT, NMIFILE, NMOFILE, NPT, NUMLADV, NVEXT, CATGOR,
CINSTSET, EXTNAM, INSTRUCTFILE, LOCINS, MCVUSE, MIFILE, MOFILE, MRKDL, NW,
PARNAM, TFILE)
```

Declarations for argument-list variables:

```
INTEGER,                                       INTENT(IN)  :: IOUNIT
INTEGER,                                       INTENT(IN)  :: LCIS
INTEGER,                                       INTENT(IN)  :: NCATS
INTEGER,                                       INTENT(IN)  :: NINSTRUCT
INTEGER,                                       INTENT(IN)  :: NMIFILE
INTEGER,                                       INTENT(IN)  :: NMOFILE
INTEGER,                                       INTENT(IN)  :: NPT
INTEGER,                                       INTENT(IN)  :: NUMLADV
INTEGER,                                       INTENT(IN)  :: NVEXT
CHARACTER (LEN=6),         DIMENSION(NMOFILE), INTENT(OUT) :: CATGOR
CHARACTER(LEN=1),          DIMENSION(LCIS),    INTENT(OUT) :: CINSTSET
CHARACTER(LEN=LENDNAM),    DIMENSION(NVEXT),   INTENT(OUT) :: EXTNAM
CHARACTER (LEN=MAX_STRING_LEN), DIMENSION(NMOFILE), INTENT(OUT) :: INSTRUCTFILE
INTEGER,                   DIMENSION(NINSTRUCT), INTENT(OUT) :: LOCINS
LOGICAL,                   DIMENSION(NCATS),   INTENT(OUT) :: MCVUSE
CHARACTER(LEN=MAX_STRING_LEN), DIMENSION(NMIFILE), INTENT(OUT) :: MIFILE
CHARACTER (LEN=MAX_STRING_LEN), DIMENSION(NMOFILE), INTENT(OUT) :: MOFILE
CHARACTER (LEN=1),         DIMENSION(NMOFILE), INTENT(OUT) :: MRKDL
INTEGER,                   DIMENSION(NPT),     INTENT(OUT) :: NW
CHARACTER(LEN=12),         DIMENSION(NPT),     INTENT(OUT) :: PARNAM
CHARACTER(LEN=MAX_STRING_LEN), DIMENSION(NMIFILE), INTENT(OUT) :: TFILE
```

Explanation of arguments:
  IOUNIT is unit number on which the "jdispatch.rdy" signal file is open and has been partially
    read. See note below.
  LCIS is equal to LENCIS of the Model Input-Output Module.
  NCATS is the number of model-calculated value categories.
  NINSTRUCT is equal to NINSTR of the Model Input-Output Module.
  NMIFILE is equal to NUMINFILE of the Model Input-Output Module.
  NMOFILE is equal to NUMOUTFILE of the Model Input-Output Module.
  NPT the number of parameters to be substituted into model-input file(s).
  NUMLADV is equal to NUML of the Model Input-Output Module.
  NVEXT is the number of model-calculated values to be extracted from model-output file(s).
  CATGOR contains the model-calculated-value category for each model-output file.
  CINSTSET is equal to CISET of the Model Input-Output Module.
  EXTNAM contains the names of values to be extracted from model-output file(s).
  INSTRUCTFILE contains the names of instruction files stored in the INSFILE array of the
    Model Input-Output Module.
  LOCINS is equal to the LCINS array of the Model Input-Output Module.
  MCVUSE indicates, for each model-calculated-value category supported by the application,
    whether extractions should be performed on model-output files associated with the
    category.

240

MIFILE is equal to MODINFILE of the Model Input-Output Module.

MOFILE is equal to MODOUTFILE of the Model Input-Output Module.

MRKDL is equal to MRKDEL of the Model Input-Output Module.

NW contains the number of characters available for writing each parameter to model-input file(s).

PARNAM contains the parameter names.

TFILE contains the names of template file(s); it is equal to the TEMPFILE array of the Model Input-Output Module.

Note:

If the "jdispatch.rdy" signal file was found by PLL_INI_RUNNER_DIM, that subroutine opens the file on unit IOUNIT, reads some of the data in it, and returns, leaving it open. A subsequent call to PLL_INI_RUNNER_POP should use the same unit number for the IOUNIT argument of PLL_INI_RUNNER_POP so that the remainder of the data in "jdispatch.rdy" can be read by PLL_INI_RUNNER_POP.

Public Subroutine – Adapt Parameter Values (ADA)

Subroutine PLL_ADA

Description: This subroutine writes model-input files. It is intended to be called from a runner program.

Argument list:

```
(NPT, NOPNT, PARNAM, PRECISPRO, NW, PVAL)
```

Declarations for argument-list variables:

```
INTEGER,                                INTENT(IN)    :: NPT
INTEGER,                                INTENT(IN)    :: NOPNT
CHARACTER(LEN=12), DIMENSION(NPT),      INTENT(IN)    :: PARNAM
INTEGER,                                INTENT(IN)    :: PRECISPRO
INTEGER,           DIMENSION(NPT),      INTENT(INOUT) :: NW
DOUBLE PRECISION,  DIMENSION(NPT),      INTENT(INOUT) :: PVAL
```

Explanation of arguments:

NPT the number of parameters to be substituted into model-input file(s).

NOPNT is the decimal point protocol setting of the Model Input-Output Module.

PARNAM contains the parameter names.

PRECISPRO is equal to PRECPR of the Model Input-Output Module. Its value is either 0 for single precision or not 0 for double precision.

NW contains the number of characters available for writing each parameter to model-input file(s).

PVAL contains the parameter values.

Public Subroutine – Execute Model (EXE)

Subroutine PLL_EXE

Description: This subroutine starts a model run. It is intended to be called from a runner program.

Argument list:

```
(IRUNRUNNER, NRUNRUNNER)
```

Declarations for argument-list variables:
```
INTEGER, INTENT(IN) :: IRUNRUNNER
INTEGER, INTENT(IN) :: NRUNRUNNER
```

Explanation of arguments:
>   IRUNRUNNER is the run number associated with the model run.
>   NRUNRUNNER is the number of model runs being made in parallel.

Note:
>   IRUNRUNNER and NRUNRUNNER are provided to the runner program in signal file "jdispar.rdy".

Public Subroutine – Extract Model-Calculated Values (EXT)

Subroutine PLL_EXT

Description: This subroutine extracts model-calculated values of dependents. It is intended to be called from a runner program.

Argument list:
```
(NCATS, NVEXT, EXTNAM, MCVUSE, EXTVAL)
```

Declarations for argument-list variables:
```
INTEGER,                               INTENT(IN)  :: NCATS
INTEGER,                               INTENT(IN)  :: NVEXT
CHARACTER (LEN=*), DIMENSION(NVEXT),   INTENT(IN)  :: EXTNAM
LOGICAL,           DIMENSION(NCATS),   INTENT(IN)  :: MCVUSE
DOUBLE PRECISION,  DIMENSION(NVEXT),   INTENT(OUT) :: EXTVAL
```

Explanation of arguments:
>   NCATS is the number of model-calculated value categories.
>   NVEXT is the number of model-calculated values to be extracted from model-output file(s).
>   EXTNAM contains the names of values to be extracted from model-output file(s).
>   MCVUSE indicates, for each model-calculated-value category supported by the application, whether extractions should be performed on model-output files associated with the category.
>   EXTVAL is populated with values of model-calculated dependents extracted from model-output file(s).

Public Subroutine – Cleanup (CLN)

Subroutine PLL_CLN

Description: This subroutine deallocates all arrays of the Parallel-Processing Module.
Argument list: Empty

Public Subroutines – Utilities

This section documents subroutines that provide general functionality needed parallelization. The subroutines are presented alphabetically as follows:
>   Subroutine PLL_MAKE_RUNS
>   Subroutine PLL_READ_DISPAR

242

Appendix J: Parallel-Processing (PLL) Module of Chapter 12

       Subroutine PLL_RUNNER_STOP
       Subroutine PLL_STOP_RUNNERS
       Subroutine PLL_WAIT
       Subroutine PLL_WRITE_RUNDEP

Subroutine PLL_MAKE_RUNS

Description: This subroutine coordinates the passing of parameter and model-calculated dependent values to and from runner directories, thus coordinating the adaptation, execution, and extraction tasks, which are performed by runners. This subroutine is intended to be called from a dispatcher program.

Argument list:

```
(LCIS, NCATS, NINSTRUCT, NMIFILE, NMOFILE, NOPNT, NPT, NUMLADV, NRUNSPLL, NVEXT,
PRECISPRO, CATGOR, CINSTSET, EXTNAM, INSTRUCTFILE, LOCINS, MCVUSE, MIFILE,
MOFILE, MRKDL, NW, PARNAM, PARVALSETS, TFILE, DEPVALSETS, COMMANDPLL)
```

Declarations for argument-list variables:

```
INTEGER,                                                INTENT(IN)  :: LCIS
INTEGER,                                                INTENT(IN)  :: NCATS
INTEGER,                                                INTENT(IN)  :: NINSTRUCT
INTEGER,                                                INTENT(IN)  :: NMIFILE
INTEGER,                                                INTENT(IN)  :: NMOFILE
INTEGER,                                                INTENT(IN)  :: NOPNT
INTEGER,                                                INTENT(IN)  :: NPT
INTEGER,                                                INTENT(IN)  :: NUMLADV
INTEGER,                                                INTENT(IN)  :: NRUNSPLL
INTEGER,                                                INTENT(IN)  :: NVEXT
INTEGER,                                                INTENT(IN)  :: PRECISPRO
CHARACTER (LEN=6),          DIMENSION(NMOFILE),         INTENT(IN)  :: CATGOR
CHARACTER(LEN=1),           DIMENSION(LCIS),            INTENT(IN)  :: CINSTSET
CHARACTER(LEN=LENDNAM),     DIMENSION(NVEXT),           INTENT(IN)  :: EXTNAM
CHARACTER (LEN=MAX_STRING_LEN), DIMENSION(NMOFILE),     INTENT(IN)  :: INSTRUCTFILE
INTEGER,                    DIMENSION(NINSTRUCT),       INTENT(IN)  :: LOCINS
LOGICAL,                    DIMENSION(NCATS),           INTENT(IN)  :: MCVUSE
CHARACTER(LEN=MAX_STRING_LEN),  DIMENSION(NMIFILE),     INTENT(IN)  :: MIFILE
CHARACTER (LEN=MAX_STRING_LEN), DIMENSION(NMOFILE),     INTENT(IN)  :: MOFILE
CHARACTER (LEN=1),          DIMENSION(NMOFILE),         INTENT(IN)  :: MRKDL
INTEGER,                    DIMENSION(NPT),             INTENT(IN)  :: NW
CHARACTER(LEN=12),          DIMENSION(NPT),             INTENT(IN)  :: PARNAM
DOUBLE PRECISION,           DIMENSION(NPT,NRUNSPLL),    INTENT(IN)  :: PARVALSETS
CHARACTER(LEN=MAX_STRING_LEN),  DIMENSION(NMIFILE),     INTENT(IN)  :: TFILE
DOUBLE PRECISION,           DIMENSION(NVEXT,NRUNSPLL),  INTENT(OUT) :: DEPVALSETS
CHARACTER(LEN=MAX_STRING_LEN),  OPTIONAL,               INTENT(IN)  :: COMMANDPLL
```

Explanation of arguments:
    LCIS is equal to LENCIS of the Model Input-Output Module.
    NCATS is the number of model-calculated value categories.
    NINSTRUCT is equal to NINSTR of the Model Input-Output Module.
    NMIFILE is equal to NUMINFILE of the Model Input-Output Module.
    NMOFILE is equal to NUMOUTFILE of the Model Input-Output Module.
    NOPNT is the decimal point protocol setting of the Model Input-Output Module.
    NPT the number of parameters to be substituted into model-input file(s).
    NUMLADV is equal to NUML of the Model Input-Output Module.

243

NRUNSPLL is the number of runs to be made in parallel.

NVEXT is the number of model-calculated values to be extracted from model-output file(s).

PRECISPRO is equal to PRECPR of the Model Input-Output Module. Its value is either 0 for single precision or not 0 for double precision.

CATGOR contains the model-calculated-value category for each model-output file.

CINSTSET is equal to CISET of the Model Input-Output Module.

EXTNAM contains the names of values to be extracted from model-output file(s).

INSTRUCTFILE contains the names of instruction files stored in the INSFILE array of the Model Input-Output Module.

LOCINS is equal to the LCINS array of the Model Input-Output Module.

MCVUSE indicates, for each model-calculated-value category supported by the application, whether extractions should be performed on model-output files associated with the category.

MIFILE is equal to MODINFILE of the Model Input-Output Module.

MOFILE is equal to MODOUTFILE of the Model Input-Output Module.

MRKDL is equal to MRKDEL of the Model Input-Output Module.

NW contains the number of characters available for writing each parameter to model-input file(s).

PARNAM contains the parameter names.

PARVALSETS contains a set of values for all parameters to be substituted into model-input file(s) for all NRUNSPLL model runs to be made.

TFILE contains the names of template file(s); it is equal to the TEMPFILE array of the Model Input-Output Module.

DEPVALSETS is populated with a set of dependent values extracted from model-output file(s) for all NRUNSPLL model runs.

COMMANDPLL is the operating-system command to initiate a model run.

Subroutine PLL_READ_DISPAR

Description: This subroutine reads the contents of a "jdispatch.rdy" file and deletes it. It is intended to be called by a runner program.

Argument list:
```
(IFAIL, NPT, IRUNRUNNER, NRUNRUNNER, PVAL)
```

Declarations for argument-list variables:
```
INTEGER,                             INTENT(OUT) :: IFAIL
INTEGER,                             INTENT(IN)  :: NPT
INTEGER,                             INTENT(OUT) :: IRUNRUNNER
INTEGER,                             INTENT(OUT) :: NRUNRUNNER
DOUBLE PRECISION, DIMENSION(NPT),    INTENT(OUT) :: PVAL
```

Explanation of arguments:
IFAIL is an error flag. IFAIL is set to 0 if no error is encountered. IFAIL is set to 1 if an error is encountered.

NPT the number of parameters to be substituted into model-input file(s).

IRUNRUNNER is the run number associated with the model run.

NRUNRUNNER is the number of model runs being made in parallel.

PVAL contains the parameter values.

Subroutine PLL_RUNNER_STOP

Description: This subroutine deletes the "jrunner.rdy", "jrundep.rdy", "jrunfail.fin", and "jdispatch.fin" files, if any of them exist. It then creates a "jrunner.fin" file and stops program execution. This subroutine is intended to be called by a runner program.
Argument list:
```
(IFLAG)
```

Declarations for argument-list variable:
```
INTEGER, OPTIONAL, INTENT(IN) :: IFLAG
```

Explanation of arguments:
   IFLAG is a flag that indicates an error status. If IFLAG equals 1, a "jrunfail.fin" signal file is created, indicating that the runner is stopping due to an error.

Subroutine PLL_STOP_RUNNERS

Description: This subroutine writes a "jdispatch.fin" file in the directories of all active runners. It is intended to be called by a dispatcher program.
Argument list: Empty

Subroutine PLL_WAIT

Description: This subroutine executes a loop continuously until a specified amount of time, in seconds, has elapsed. When this amount of time has elapsed, control is returned to the calling program unit.
Argument list:
```
(TSECS)
```

Declaration for argument-list variable:
```
DOUBLE PRECISION, OPTIONAL, INTENT(IN) :: TSECS
```

Explanation of argument:
 TSECS is the amount of time, in seconds, that is to elapse before control is returned to the calling program unit.
Note:
 If TSECS is not present, the value of the module variable WAIT is used as the amount of time, in seconds, before control is returned to the calling program unit.

Subroutine PLL_WRITE_RUNDEP

Description: This subroutine writes the run number and values of model-calculated dependents to a "jrundep.rdy" file. It is intended to be called by a runner program.
Argument list:
```
(IRUN, NVEXT, EXTVAL)
```

Declarations for argument-list variables:
```
INTEGER,                            INTENT(IN) :: IRUN
INTEGER,                            INTENT(IN) :: NVEXT
DOUBLE PRECISION, DIMENSION(NVEXT), INTENT(IN) :: EXTVAL
```

Explanation of arguments:
 IRUN is the run number.
 NVEXT is the number of model-calculated values to be extracted from model-output file(s).
 EXTVAL contains values of model-calculated dependents extracted from model-output file(s).

Appendix K: Sensitivity (SEN) Module of Chapter 13

Private Data

Declaration of variables:
```
INTEGER,                 ALLOCATABLE, DIMENSION(:) :: IDONE
INTEGER                                            :: ISENPERT
DOUBLE PRECISION,        ALLOCATABLE, DIMENSION(:) :: DELB
DOUBLE PRECISION,        ALLOCATABLE, DIMENSION(:) :: PVALFWD
INTEGER                                            :: NPARDONE
CHARACTER(LEN=MAX_STRING_LEN)                      :: DI_DERFILE
INTEGER                                            :: DI_NSKIP
INTEGER                                            :: DI_NDEP
INTEGER                                            :: DI_NPAR
CHARACTER(LEN=10)                                  :: DI_ORIENTATION
CHARACTER(LEN=MAX_STRING_LEN)                      :: DI_DERFORMAT
CHARACTER(LEN=12),       ALLOCATABLE, DIMENSION(:) :: DI_PARNAM
CHARACTER(LEN=LENDNAM),  ALLOCATABLE, DIMENSION(:) :: DI_DEPNAM
INTEGER,                 ALLOCATABLE, DIMENSION(:) :: DI_XROW
INTEGER,                 ALLOCATABLE, DIMENSION(:) :: DI_XCOL
```

Explanation of variables:
IDONE contains a flag for each row of sensitivity matrix, indicating if the row has been populated for the corresponding parameter.

ISENPERT is a flag that indicates the current method for generating sensitivities. 0 indicates model-calculated sensitivities; 1 indicates perturbation sensitivities.

DELB contains the perturbation amount for each adjustable parameter.

PVALFWD contains the forward-perturbed value for each adjustable parameter.

NPARDONE is the number of parameters for which sensitivities have been calculated.

DI_DERFILE is the name of the model-generated file containing model-calculated derivatives of dependents with respect to parameters.

DI_NSKIP is the number of lines at the top of file DI_DERFILE to be skipped before the first line of model-calculated derivatives.

DI_NDEP is the number of dependents for which model-calculated derivatives are to be read from file DI_DERFILE.

DI_NPAR is the number of parameters for which model-calculated derivatives are to be read from file DI_DERFILE.

DI_ORIENTATION is a flag indicating the orientation of the 2-D array of derivatives in file DI_DERFILE. Two values of DI_ORIENTATION are valid: "ROW/DEP" indicates that each row contains derivatives for one dependent; "ROW/PAR" indicates that each row contains derivatives for one parameter.

DI_DERFORMAT is the Fortran format, or "(FREE)", to be used to read each row of derivatives in file DI_DEREFILE. If DI_DERFORMAT is "(FREE)", each row is read in free format.

DI_PARNAM contains the parameter names, as ordered in file DI_DERFILE.

DI_DEPNAM contains the dependent names, as ordered in file DI_DERFILE.

DI_XROW contains the number of the row in the sensitivity matrix corresponding to each parameter in DI_PARNAM.

DI_XCOL contains the number of the column in the sensitivity matrix corresponding to each dependent in DI_DEPNAM.

Appendix K: Sensitivity (SEN) Module of Chapter 13

Public Subroutine – Initialize (INI)

Subroutine SEN_INI

Description: This subroutine allocates memory for forward- and backward-difference model-calculated values and for perturbation amount for each parameter. The DELB, IDONE, and PVALFWD module arrays are allocated, and the Derivatives Interface file is read.

Argument list:
```
(DERIV_INTERFACE, IOUT, NDEP, NPE, NPT, IPTR, ISENMETHOD, PARNAM, DEPNAM, IFAIL)
```

Declarations for argument-list variables:
```
CHARACTER(LEN=MAX_STRING_LEN),              INTENT(IN)  :: DERIV_INTERFACE
INTEGER,                                    INTENT(IN)  :: IOUT
INTEGER,                                    INTENT(IN)  :: NDEP
INTEGER,                                    INTENT(IN)  :: NPE
INTEGER,                                    INTENT(IN)  :: NPT
INTEGER, DIMENSION(NPE),                    INTENT(IN)  :: IPTR
INTEGER, DIMENSION(NPT),                    INTENT(IN)  :: ISENMETHOD
CHARACTER(LEN=12), DIMENSION(NPT),          INTENT(IN)  :: PARNAM
CHARACTER(LEN=LENDNAM), DIMENSION(NDEP),    INTENT(IN)  :: DEPNAM
INTEGER,                                    INTENT(OUT) :: IFAIL
```

Explanation of arguments:
DERIV_INTERFACE is the name of a Derivatives Interface file, which will be used to define the protocol for reading model-calculated derivatives.
IOUT is the unit number associated with the file to which output is to be written.
NDEP is the number of dependent variables.
NPE is the number of adjustable parameters.
NPT is the number of parameters.
IPTR contains, for each adjustable parameter, the position (element number) in the ISENMETHOD and PARNAM arrays corresponding to the adjustable parameter.
ISENMETHOD contains a flag that identifies the sensitivity method used for each parameter. The possible values have the following meanings:
ISENMETHOD = 0 – Sensitivities are to be obtained from model output.
ISENMETHOD = 1 – Sensitivities are to be obtained by forward-difference perturbation.
ISENMETHOD = 2 – Sensitivities are to be obtained by central-difference perturbation (two-point method).
PARNAM contains the names of the parameters.
DEPNAM contains the names of the dependents.
IFAIL is an error flag. IFAIL is set to 0 if no error is encountered. IFAIL is set to 1 if an error is encountered.

Public Subroutine – Define Job (DEF)

Subroutine SEN_DEF

Description: This subroutine assigns values to variables that control the functioning of the Sensitivity Module. The SENSDONE argument is set to false.

Appendix K: Sensitivity (SEN) Module of Chapter 13

Argument list:
```
(CTRLJOB, IOUT, NDEP, NPE, NPT, IPTR, ISENMETHOD, PADJ, PARNAM, SENSDONE, X,
ISNPRT, NUMPPL)
```

Declarations for argument-list variables:
```
CHARACTER(LEN=*),                        INTENT(IN)    :: CTRLJOB
INTEGER,                                 INTENT(IN)    :: IOUT
INTEGER,                                 INTENT(IN)    :: NDEP
INTEGER,                                 INTENT(IN)    :: NPE
INTEGER,                                 INTENT(IN)    :: NPT
INTEGER, DIMENSION(NPE),                 INTENT(IN)    :: IPTR
INTEGER, DIMENSION(NPT),                 INTENT(IN)    :: ISENMETHOD
LOGICAL, DIMENSION(NPT),                 INTENT(IN)    :: PADJ
CHARACTER(LEN=12), DIMENSION(NPT),       INTENT(IN)    :: PARNAM
LOGICAL,                                 INTENT(INOUT) :: SENSDONE
DOUBLE PRECISION, DIMENSION(NPE,NDEP),   INTENT(INOUT) :: X
INTEGER,                                 INTENT(OUT)   :: ISNPRT
INTEGER,                                 INTENT(OUT)   :: NUMPPL
```

Explanation of arguments:
> CTRLJOB is the text identifying the job of the current iteration of the Control Loop.
> IOUT is the unit number associated with the file to which output is to be written.
> NDEP is the number of dependent variables.
> NPE is the number of adjustable parameters.
> NPT is the number of parameters.
> IPTR contains, for each adjustable parameter, the position (element number) in the ISENMETHOD, PADJ, and PARNAM arrays corresponding to the adjustable parameter.
> ISENMETHOD contains a flag that identifies the sensitivity method used for each parameter. The possible values have the following meanings:
> ISENMETHOD = 0 – Sensitivities are to be obtained from model output.
> ISENMETHOD = 1 – Sensitivities are to be obtained by forward-difference perturbation.
> ISENMETHOD = 2 – Sensitivities are to be obtained by central-difference perturbation (two-point method).
> PADJ indicates whether each parameter is adjustable. If PADJ(I) is true, parameter I is adjustable.
> PARNAM contains the parameter names.
> SENSDONE is used to communicate to the calling routine whether or not the Sensitivity Module is ready to relinquish control.
> X contains the dependent-variable sensitivities to each parameter.
> ISNPRT is a flag indicating the method by which sensitivities are to be calculated in the current iteration of the Control Loop. It will have one of the following values:
> 0 – Model-calculated sensitivity
> 1 – Perturbation sensitivity
> NUMPPL is the number of iterations of the PPL loop required for the current iteration of the Control loop, based on the ISENMETHOD value for each adjustable parameter.

Note:
> The method by which sensitivities are to be calculated in the current iteration of the Control Loop is written to IOUT. The method is one of: "MODEL-CALCULATED SENSITIVITY" or "PERTURBATION SENSITIVITY".

Appendix K: Sensitivity (SEN) Module of Chapter 13

Public Subroutine – Generate Parameter Values (GEN)

Subroutine SEN_GEN

Description: This subroutine generates a set of parameter values to be used in the calculation of sensitivities.

Argument list:
```
(CTRLJOB, KPPL, NPE, NPT, IOUT, BINC, IPTR, ISENMETHOD, NOPNT, NW, PARNAM,
PRECPR, PVAL, KPE, PSETTMP)
```

Declarations for argument-list variables:
```
CHARACTER(LEN=*),                       INTENT(IN)  :: CTRLJOB
INTEGER,                                INTENT(IN)  :: KPPL
INTEGER,                                INTENT(IN)  :: NPE
INTEGER,                                INTENT(IN)  :: NPT
INTEGER,                                INTENT(IN)  :: IOUT
DOUBLE PRECISION,    DIMENSION(NPT),    INTENT(IN)  :: BINC
INTEGER,             DIMENSION(NPE),    INTENT(IN)  :: IPTR
INTEGER,             DIMENSION(NPT),    INTENT(IN)  :: ISENMETHOD
INTEGER,                                INTENT(IN)  :: NOPNT
INTEGER,             DIMENSION(NPT),    INTENT(IN)  :: NW
CHARACTER(LEN=*),    DIMENSION(NPT),    INTENT(IN)  :: PARNAM
INTEGER,                                INTENT(IN)  :: PRECPR
DOUBLE PRECISION,    DIMENSION(NPT),    INTENT(IN)  :: PVAL
INTEGER,                                INTENT(OUT) :: KPE
DOUBLE PRECISION,    DIMENSION(NPT),    INTENT(OUT) :: PSETTMP
```

Explanation of arguments:

CTRLJOB is the text identifying the job of the current iteration of the Control Loop. If CTRLJOB is "FORWARD&SENS", the subroutine recognizes that the model run will generate both dependents and model-calculated sensitivities.

KPPL is the iteration counter for the PPL loop.

NPE is the number of adjustable parameters.

NPT is the number of parameters.

IOUT is the unit number associated with the file to which output is to be written.

BINC contains the perturbation increment for each parameter. BINC is added to the current parameter value to effect a forward perturbation. Also, when calculating central-difference sensitivities BINC is subtracted from the currrent parameter value to effect a backward perturbation.

IPTR contains, for each adjustable parameter, the position (element number) in the ISENMETHOD, NW, PARNAM, PVAL, and PVALTMP arrays corresponding to the adjustable parameter.

ISENMETHOD contains a flag that identifies the sensitivity method used for each parameter. The possible values have the following meanings:

ISENMETHOD = 0 – Sensitivities are to be obtained from model output.

ISENMETHOD = 1 – Sensitivities are to be obtained by forward-difference perturbation.

ISENMETHOD = 2 – Sensitivities are to be obtained by central-difference perturbation (two-point method).

NOPNT is a flag indicating the decimal point protocol for writing parameter values.

NOPNT=0 indicates that a decimal point should be included when writing parameter values to a model-input file, and NOPNT=1 indicates that a decimal point may be omitted.

NW contains the number of characters available for writing each parameter to model-input files.

PARNAM contains the parameter names.

PRECPR is a flag indicating the precision protocol for writing parameter values. Valid values for PRECPR and their meanings are:

PRECPR = 0 – values will be written in single precision, with exponent in "E" format when required.

PRECPR = 1 – values will be written in double precision, with exponent in "E" format when required.

PRECPR = 2 – values will be written in double precision, with exponent in "D" format when required.

PVAL contains current (unperturbed) parameter values.

KPE is the sequential number (among the adjustable parameters) of the parameter currently being perturbed; it can have a value in the range 1 to NPE.

PSETTMP is populated with a set of parameter values, including a perturbed value, for use in generating dependent-variable values for calculating sensitivities.

Public Subroutine – Execute Model (EXE)

Subroutine SEN_EXE_SELECT

Description: This subroutine determines the number of a model command line to be executed to make a model run that will generate dependent-variable sensitivities.

Argument list:
```
(IOUT, ICOMMAND, COMMAND)
```

Declarations for argument-list variables:
```
INTEGER,                                    INTENT(IN)  :: IOUT
INTEGER,                                    INTENT(OUT) :: ICOMMAND
CHARACTER(LEN=MAX_STRING_LEN), OPTIONAL, INTENT(OUT) :: COMMAND
```

Explanation of arguments:

IOUT is the unit number associated with the file to which output is to be written.

ICOMMAND is the command number (position in list of commands stored in MODCOMLINE).

COMMAND is the command text associated with command ICOMMAND.

Public Subroutines – Use Extracted Values (UEV)

This section describes subroutines used to manipulate values that have been extracted from model output files. The subroutines are presented in alphabetical order as follows:

Subroutine SEN_UEV_DX_READ_MATRIX
Subroutine SEN_UEV_DX_READ_SU
Subroutine SEN_UEV_DX_WRITE_MATRIX
Subroutine SEN_UEV_DX_WRITE_SEN
Subroutine SEN_UEV_LNX
Subroutine SEN_UEV_POPX_MODCALC
Subroutine SEN_UEV_POPXROW_DIFF
Subroutine SEN_UEV_WRITESENTABLE

Appendix K: Sensitivity (SEN) Module of Chapter 13

Subroutine SEN_UEV_DX_READ_MATRIX

Description: This subroutine reads one data-exchange file, which may be of type _s1, _sd, or _su (Table 17-4).

Argument list:
```
(IUSM, NDEP, NPE, X, COLNAMES, ROWNAMES)
```

Declarations for argument-list variables:
```
INTEGER,                                          INTENT(IN)    :: IUSM
INTEGER,                                          INTENT(IN)    :: NDEP
INTEGER,                                          INTENT(IN)    :: NPE
DOUBLE PRECISION, DIMENSION(NPE,NDEP),            INTENT(OUT)   :: X
CHARACTER(LEN=*), DIMENSION(NPE),  OPTIONAL, INTENT(INOUT) :: COLNAMES
CHARACTER(LEN=*), DIMENSION(NDEP), OPTIONAL, INTENT(INOUT) :: ROWNAMES
```

Explanation of arguments:

IUSM is the unit number on which a data-exchange file of type _s1, _sd, or _su has been opened for reading.

NDEP is the number of dependents.

NPE is the number of adjustable parameters.

X is populated with the sensitivity matrix read from unit IUSM.

COLNAMES, if present, is populated with column names read from the data-exchange file. See note 2.

ROWNAMES, if present, is populated with row names read from the data-exchange file. See note 2.

Notes:

1. If the file associated with unit IUSM is of type _s1, the sensitivities contained in the file are expected to be 1-percent scaled sensitivities. If the file is of type _sd, the sensitivities are expected to be dimensionless scaled sensitivities. If the file is of type _su, the sensitivities are expected to be unscaled sensitivities.

2. Both optional arguments COLNAMES and ROWNAMES must be either present or absent; if only one is present, an error message is written to the screen and program execution is halted.

Subroutine SEN_UEV_DX_READ_SU

Description: This subroutine reads unscaled sensitivities of dependents with respect to parameters from a data-exchange file of type _su (Table 17-4).

Argument list:
```
(NDEP, NPE, OBSNAM, OUTNAM, PARNAM, PLOTSYMBOL, XSENS)
```

Declarations for argument-list variables:
```
INTEGER,                                  INTENT(IN)  :: NDEP
INTEGER,                                  INTENT(IN)  :: NPE
CHARACTER(LEN=MAX_STRING_LEN),            INTENT(IN)  :: OUTNAM
CHARACTER(LEN=LENDNAM), DIMENSION(NDEP),  INTENT(OUT) :: OBSNAM
CHARACTER(LEN=12),      DIMENSION(NPE),   INTENT(OUT) :: PARNAM
INTEGER,                DIMENSION(NDEP),  INTENT(OUT) :: PLOTSYMBOL
DOUBLE PRECISION,   DIMENSION(NPE,NDEP),  INTENT(OUT) :: XSENS
```

Appendix K: Sensitivity (SEN) Module of Chapter 13

Explanation of arguments:

NDEP is the number of dependents.

NPE is the number of parameters.

OUTNAM is the root name for the data-exchange file.

OBSNAM is populated with the dependent names read from the _su file.

PARNAM is populated with the parameter names read from the _su file.

PLOTSYMBOL is populated with the plot-symbol values read from the _su file.

XSENS is populated with the unscaled sensitivities read from the _su file.

Subroutine SEN_UEV_DX_WRITE_MATRIX

Description: This subroutine writes arrays as needed to produce files such as the data-exchange files with filename extensions "_s1", "_sd", or "_su" (Table 17-4).

Argument list:

```
(IUSM, NDEP, NPE, NPT, OBSNAM, IPTR, PARNAM, X, IPLOTOPT)
```

Declarations for argument-list variables:

```
INTEGER,                                      INTENT(IN) :: IUSM
INTEGER,                                      INTENT(IN) :: NDEP
INTEGER,                                      INTENT(IN) :: NPE
INTEGER,                                      INTENT(IN) :: NPT
CHARACTER(LEN=LENDNAM), DIMENSION(NDEP), INTENT(IN) :: OBSNAM
INTEGER,                DIMENSION(NPE),  INTENT(IN) :: IPTR
CHARACTER(LEN=12),      DIMENSION(NPT),  INTENT(IN) :: PARNAM
DOUBLE PRECISION,    DIMENSION(NPE,NDEP), INTENT(IN) :: X
INTEGER, OPTIONAL,      DIMENSION(NDEP), INTENT(IN) :: IPLOTOPT
```

Explanation of arguments:

IUSM is the unit number on which a data-exchange file of type _s1, _sd or _su has been opened for writing.

NDEP is the number of dependents.

NPE is the number of adjustable parameters and the number of rows in the sensitivity matrix to be written.

NPT is the total number of parameters.

OBSNAM contains the dependent names.

IPTR contains, for each adjustable parameter, the position (element number) in the PARNAM array corresponding to the adjustable parameter.

PARNAM contains the parameter names.

X contains the sensitivities of dependents with respect to adjustable parameters.

IPLOTOPT, if present, contains plot-symbol values to be written to the data-exchange file. If IPLOTOPT is absent, SEN_UEV_DX_WRITE_MATRIX will use plot-symbol values returned by DEP_GET_PLOTSYMBOL, which are the values stored in the PLOTSYMBOL array of the Dependents Module.

Notes:

1. The sensitivities in X are written to unit IUSM without modification. If IUSM is associated with a file with extension "_s1", X is expected to contain 1-percent scaled sensitivities. If IUSM is associated with a file with extension "_sd", X is expected to contain dimensionless scaled sensitivities. If IUSM is associated with a file with extension "_su", X is expected to contain unscaled sensitivities.

Appendix K: Sensitivity (SEN) Module of Chapter 13

Subroutine SEN_UEV_DX_WRITE_SEN

Description: This subroutines writes five types of sensitivities of simulated equivalents to observations with respect to adjustable parameters to data-exchange files of type: _su, _sd, _so, _sc, and _s1 (Table 17-4).

Argument list:
```
(NCOVMAT, NOBS, NPE, NPS, LN, OBSNAM, OUTNAM, IPTR, PARNAM, PVAL, WTFULLSQR,
XSENS)
```

Declarations for argument-list variables:
```
INTEGER,                                    INTENT(IN) :: NCOVMAT
INTEGER,                                    INTENT(IN) :: NOBS
INTEGER,                                    INTENT(IN) :: NPE
INTEGER,                                    INTENT(IN) :: NPS
INTEGER,                    DIMENSION(NPS), INTENT(IN) :: LN
CHARACTER(LEN=LENDNAM), DIMENSION(NOBS), INTENT(IN) :: OBSNAM
CHARACTER(LEN=MAX_STRING_LEN),           INTENT(IN) :: OUTNAM
INTEGER,                    DIMENSION(NPE), INTENT(IN) :: IPTR
CHARACTER(LEN=12),          DIMENSION(NPS), INTENT(IN) :: PARNAM
DOUBLE PRECISION,           DIMENSION(NPS), INTENT(IN) :: PVAL
TYPE (CDMATRIX),                            INTENT(IN) :: WTFULLSQR
DOUBLE PRECISION,    DIMENSION(NPE,NOBS), INTENT(IN) :: XSENS
```

Explanation of arguments:
   NCOVMAT is the number of variance-covariance matrices for groups of correlated observations and prior-information equations.
   NOBS is the number of observations.
   NPE is the number of adjustable parameters.
   NPS is the total number of parameters.
   LN contains, for each parameter, a flag indicating if the parameter is log-transformed for analysis. If LN = 0, the parameter is not log-transformed, and if LN > 0, the parameter is log-transformed.
   OBSNAM contains the observation names.
   OUTNAM is the root name for the data-exchange files.
   IPTR contains, for each adjustable parameter, the position (element number) in the PARNAM array corresponding to the adjustable parameter.
   PARNAM contains the parameter names.
   PVAL contains the parameter values.
   WTFULLSQR contains the square root of the weight matrix for observations.
   XSENS contains the matrix of sensitivities of simulated equivalents to observations with respect to adjustable parameters.

Subroutine SEN_UEV_LNX

Description: This subroutine converts native-space derivatives to log-space derivatives (by multiplying by the parameter value) for log-transformed parameters.

Argument list:
```
(NDEP, NPE, NPT, IPTR, LN, PVAL, X, ICONVERT_DERIV)
```

Declarations for argument-list variables:
```
INTEGER,                                    INTENT(IN)    :: NDEP
```

Appendix K: Sensitivity (SEN) Module of Chapter 13

```
INTEGER,                                      INTENT(IN)      :: NPE
INTEGER,                                      INTENT(IN)      :: NPT
INTEGER,            DIMENSION(NPE),           INTENT(IN)      :: IPTR
INTEGER,            DIMENSION(NPT),           INTENT(IN)      :: LN
DOUBLE PRECISION,   DIMENSION(NPT),           INTENT(IN)      :: PVAL
DOUBLE PRECISION,   DIMENSION(NPE,NDEP),      INTENT(INOUT)   :: X
INTEGER,        DIMENSION(NPT), OPTIONAL,     INTENT(IN)      :: ICONVERT_DERIV
```

Explanation of arguments:

NDEP is the number of dependent variables.

NPE is the number of adjustable parameters.

NPT is the number of parameters.

IPTR contains, for each adjustable parameter, the position (element number) in the LN and PVAL arrays corresponding to the adjustable parameter.

LN contains, for each parameter, a flag indicating if the parameter is log-transformed for analysis. If LN = 0, the parameter is not log-transformed, and if LN > 0, the parameter is log-transformed.

PVAL contains the current parameter values.

X contains the dependent-variable sensitivities to each parameter.

ICONVERT_DERIV is a flag: 1 indicates that sensitivity should be multiplied by parameter value if it is a log-transformed parameter, 0 if it should not be multiplied because it is read as a transformed sensitivity using the Derivatives Interface file.

Subroutine SEN_UEV_POPX_MODCALC

Description: This subroutine populates rows of the sensitivity array with model-calculated derivatives, and determines if sensitivities have been calculated for all adjustable parameters.

Argument list:
```
(IOUT, NDEP, NPE, NPT, IPTR, ISENMETHOD, SENSDONE, X)
```

Declarations for argument-list variables:
```
INTEGER,                                      INTENT(IN)      :: IOUT
INTEGER,                                      INTENT(IN)      :: NDEP
INTEGER,                                      INTENT(IN)      :: NPE
INTEGER,                                      INTENT(IN)      :: NPT
INTEGER, DIMENSION(NPE),                      INTENT(IN)      :: IPTR
INTEGER, DIMENSION(NPT),                      INTENT(IN)      :: ISENMETHOD
LOGICAL,                                      INTENT(INOUT)   :: SENSDONE
DOUBLE PRECISION, DIMENSION(NPE,NDEP),        INTENT(INOUT)   :: X
```

Explanation of arguments:

IOUT is the unit number associated with the file to which output is to be written.

NDEP is the number of dependent variables.

NPE is the number of adjustable parameters.

NPT is the number of parameters.

IPTR contains, for each adjustable parameter, the position (element number) in the LN and PVAL arrays corresponding to the adjustable parameter.

ISENMETHOD contains a flag that identifies the sensitivity method used for each parameter. The possible values have the following meanings:

ISENMETHOD = 0 – Sensitivities are to be obtained from model output.

ISENMETHOD = 1 – Sensitivities are to be obtained by forward-difference perturbation.

Appendix K: Sensitivity (SEN) Module of Chapter 13

ISENMETHOD = 2 – Sensitivities are to be obtained by central-difference perturbation (two-point method).

SENSDONE is used to communicate to the calling routine whether or not the Sensitivity Module is ready to relinquish control. It is set to true if sensitivities have been calculated for all adjustable parameters.

X contains the dependent-variable sensitivities to each parameter.

Note:

Rows in the X array corresponding to adjustable parameters listed in the Derivatives Interface File are populated with sensitivities obtained from the model-output file identified as DERFILE in the Derivatives Interface file (see the "Derivatives Interface File" section of Appendix A).

Subroutine SEN_UEV_POPXROW_DIFF

Description: This subroutine populates a row of the sensitivity array with derivatives calculated by finite-differences. Two sets of model-calculated dependent values are assumed to have been generated by two model runs that differ by a perturbation of one parameter. It is necessary to invoke SEN_GEN prior to invoking SEN_UEV_POPXROW_DIFF, to enable the Sensitivity Module to calculate and store differences in parameter values for each perturbed parameter.

Argument list:
```
(KPE, NDEP, SIMEQ1, SIMEQ2, XROW)
```

Declarations for argument-list variables:
```
INTEGER,                              INTENT(IN)    :: KPE
INTEGER,                              INTENT(IN)    :: NDEP
DOUBLE PRECISION, DIMENSION(NDEP),    INTENT(IN)    :: SIMEQ1
DOUBLE PRECISION, DIMENSION(NDEP),    INTENT(IN)    :: SIMEQ2
DOUBLE PRECISION, DIMENSION(NDEP),    INTENT(INOUT) :: XROW
```

Explanation of arguments:

KPE is the number of the perturbed parameter, between 1 and the number of adjustable parameters.

NDEP is the number of dependents.

SIMEQ1 contains the dependent values resulting from a model run where the parameter had value P1. See note.

SIMEQ2 contains the dependent values resulting from a model run where the parameter had value P2. See note.

XROW is the array to be populated with sensitivities.

Note:

The sensitivity is calculated as (SIMEQ1(j) – SIMEQ2(j)) / DELB(KPE) for each dependent, denoted by j. The differences of two values of each perturbed parameter, defined as (P_1 – P_2), is stored in the DELB array (a PRIVATE data item in the Sensitivity Module) when subroutine SEN_GEN is invoked to calculate sets of parameter values for generating perturbation sensitivities.

Appendix K: Sensitivity (SEN) Module of Chapter 13

Subroutine SEN_UEV_WRITESENTABLE

Description: This subroutine writes a formatted table of dependent-variable sensitivities, with a column for each adjustable parameter. When the number of parameters exceeds NCOLS, multiple table sections are written, with NCOLS parameters included in each section.

Argument list:
```
(NDEP, NPE, NPT, DEPNAM, HEADING, IOUT, IPTR, NCOLS, PARNAM, X)
```

Declarations for argument-list variables:
```
INTEGER,                                    INTENT(IN) :: NDEP
INTEGER,                                    INTENT(IN) :: NPE
INTEGER,                                    INTENT(IN) :: NPT
CHARACTER(LEN=LENDNAM), DIMENSION(NDEP),    INTENT(IN) :: DEPNAM
CHARACTER(LEN=*),                           INTENT(IN) :: HEADING
INTEGER,                                    INTENT(IN) :: IOUT
INTEGER, DIMENSION(NPE),                    INTENT(IN) :: IPTR
INTEGER,                                    INTENT(IN) :: NCOLS
CHARACTER(LEN=12), DIMENSION(NPT),          INTENT(IN) :: PARNAM
DOUBLE PRECISION, DIMENSION(NPE,NDEP),      INTENT(IN) :: X
```

Explanation of arguments:
NDEP is the number of dependent variables.
NPE is the number of adjustable parameters.
NPT is the number of parameters.
DEPNAM contains the names of the dependent variables (observations or predictions)
HEADING is the text to be written as the table heading.
IOUT is the unit number associated with the file to which output is to be written.
IPTR contains, for each adjustable parameter, the position (element number) in the PARNAM array corresponding to the adjustable parameter.
NCOLS is the number of parameters to be included in each table section.
PARNAM contains the parameter names.
X contains the dependent-variable sensitivities to each parameter.

Public Subroutine – Cleanup (CLN)

Subroutine SEN_CLN

Description: This subroutine deallocates all allocatable arrays that are stored in the Sensitivity Module.
Argument list: Empty

Appendix L: Statistics (STA) Module of Chapter 14

Public Data

Declaration of variables:
```
DOUBLE PRECISION, ALLOCATABLE, DIMENSION(:) :: WTDOBS
```

Explanation of variables:
    WTDOBS is an array of weighted observations.

Public Subroutine – Initialize (INI)

Subroutine STA_INI

Description: This subroutine populates the WTDOBS array of the Statistics Module with weighted
    observations.
Argument list:
```
(NOBS, OBSVAL, WTMATSQR)
```

Declarations for argument-list variables:
```
INTEGER,                             INTENT(IN) :: NOBS
DOUBLE PRECISION, DIMENSION(NOBS),   INTENT(IN) :: OBSVAL
TYPE (CDMATRIX),                     INTENT(IN) :: WTMATSQR
```

Explanation of arguments:
    NOBS is the number of observations.
    OBSVAL contains the observation values.
    WTMATSQR contains the square root of the weight matrix.

Public Subroutines – Use Extracted Values (UEV)

    This section describes subroutines used to manipulate values that have been extracted from
model output files. The subroutines are presented in alphabetical order as follows:
    Subroutine STA_UEV_DX_READ_DM
    Subroutine STA_UEV_DX_WRITE_DM
    Subroutine STA_UEV_DX_WRITE_NM
    Subroutine STA_UEV_FIT
    Subroutine STA_UEV_INIT

Subroutine STA_UEV_DX_READ_DM

Description: This subroutine reads a summary of model data from a data-exchange file of type
    _dm (**Table 17-7**).
Argument list:
```
(IFAIL, OUTNAM, AIC, BIC, DTLA, HQ, ICONVERGE, ITERP, KASHYAP, MLOFD, MLOFDP,
MODELLENGTH, MODELMASS, MODELNAME, MODELTIME, MPR, NDINC, NPE, NPERD, NPS, NOBS,
STAT1, STAT2, STDERR, VAR)
```

258

Appendix L: Statistics (STA) Module of Chapter 14

Declarations for argument-list variables:
```
INTEGER,                            INTENT(INOUT)  :: IFAIL
CHARACTER(LEN=MAX_STRING_LEN),      INTENT(IN)     :: OUTNAM
DOUBLE PRECISION,                   INTENT(OUT)    :: AIC
DOUBLE PRECISION,                   INTENT(OUT)    :: BIC
DOUBLE PRECISION,                   INTENT(OUT)    :: DTLA
DOUBLE PRECISION,                   INTENT(OUT)    :: HQ
CHARACTER(LEN=3),                   INTENT(OUT)    :: ICONVERGE
INTEGER,                            INTENT(OUT)    :: ITERP
DOUBLE PRECISION,                   INTENT(OUT)    :: KASHYAP
DOUBLE PRECISION,                   INTENT(OUT)    :: MLOFD
DOUBLE PRECISION,                   INTENT(OUT)    :: MLOFDP
CHARACTER(LEN=12),                  INTENT(OUT)    :: MODELLENGTH
CHARACTER(LEN=12),                  INTENT(OUT)    :: MODELMASS
CHARACTER(LEN=12),                  INTENT(OUT)    :: MODELNAME
CHARACTER(LEN=12),                  INTENT(OUT)    :: MODELTIME
INTEGER,                            INTENT(OUT)    :: MPR
INTEGER,                            INTENT(OUT)    :: NDINC
INTEGER,                            INTENT(OUT)    :: NPE
INTEGER,                            INTENT(OUT)    :: NPERD
INTEGER,                            INTENT(OUT)    :: NPS
INTEGER,                            INTENT(OUT)    :: NOBS
DOUBLE PRECISION,                   INTENT(OUT)    :: STAT1
DOUBLE PRECISION,                   INTENT(OUT)    :: STAT2
DOUBLE PRECISION,                   INTENT(OUT)    :: STDERR
DOUBLE PRECISION,                   INTENT(OUT)    :: VAR
```

Explanation of arguments:

IFAIL is an error flag. Any value other that 0 indicates an error has occurred.

OUTNAM is the root file name for output files.

AIC is a measure of model fit and parsimony, calculated as shown in Table 14-1.

BIC is a measure of model fit and parsimony, calculated as shown in Table 14-1.

DTLA is the determinant of the Fisher Information Matrix, calculated as shown in Table 14-1.

HQ is a measure of model fit and parsimony, calculated as shown in Table 14-1.

ICONVERGE indicates whether the regression converged ("YES"), or not ("NO")

ITERP is the number of iterations required for the regression to converge.

KASHYAP is a measure of model fit and parsimony, calculated as shown in Table 14-1.

MLOFD is the Maximum Likelihood Objective Function evaluated using the observations, calculated as shown in Table 14-1.

MLOFDP is the Maximum Likelihood Objective Function evaluated using observations and prior information, calculated as shown in Table 14-1.

MODELLENGTH is the length unit used in the model.

MODELMASS is the mass unit used in the model.

MODELNAME is the model name.

MODELTIME is the time unit used in the model.

MPR is the number of prior-information equations.

NDINC is the number of observations included in the final iteration of the parameter estimation.

NPE is the number of adjustable parameters.

NPERD is the number of parameters estimated in the iteration from which the optimal values are determined.

NPS is the total number of parameters in the parameter input blocks.

NOBS is the number of observations.

STAT1 is the correlation of weighted residuals for dependent variable observations to standard normal statistics.

STAT2 is the correlation of weighted residuals for dependent variable observations and prior-information equations to standard normal statistics.

STDERR is the standard error of the regression, $VAR^{0.5}$.

VAR is the calculated error variance, calculated as shown in Table 14-1.

Subroutine STA_UEV_DX_WRITE_DM

Description: This subroutine writes a summary of model data.

Argument list:

```
(AIC, BIC, DTLA, HQ, IFO, ITERP, KASHYAP, MLOFD, MLOFDP, MODELLENGTH, MODELMASS,
MODELNAME, MODELTIME, MPR, NDINC, NPE, NPERD, NPS, NOBS, OUTNAM, STAT1, STAT2,
VAR)
```

Declarations for argument-list variables:

```
DOUBLE PRECISION,                INTENT(IN) :: AIC
DOUBLE PRECISION,                INTENT(IN) :: BIC
DOUBLE PRECISION,                INTENT(IN) :: DTLA
DOUBLE PRECISION,                INTENT(IN) :: HQ
INTEGER,                         INTENT(IN) :: IFO
INTEGER,                         INTENT(IN) :: ITERP
DOUBLE PRECISION,                INTENT(IN) :: KASHYAP
DOUBLE PRECISION,                INTENT(IN) :: MLOFD
DOUBLE PRECISION,                INTENT(IN) :: MLOFDP
CHARACTER(LEN=12),               INTENT(IN) :: MODELLENGTH
CHARACTER(LEN=12),               INTENT(IN) :: MODELMASS
CHARACTER(LEN=12),               INTENT(IN) :: MODELNAME
CHARACTER(LEN=12),               INTENT(IN) :: MODELTIME
INTEGER,                         INTENT(IN) :: MPR
INTEGER,                         INTENT(IN) :: NDINC
INTEGER,                         INTENT(IN) :: NPE
INTEGER,                         INTENT(IN) :: NPERD
INTEGER,                         INTENT(IN) :: NPS
INTEGER,                         INTENT(IN) :: NOBS
CHARACTER(LEN=MAX_STRING_LEN),   INTENT(IN) :: OUTNAM
DOUBLE PRECISION,                INTENT(IN) :: STAT1
DOUBLE PRECISION,                INTENT(IN) :: STAT2
DOUBLE PRECISION,                INTENT(IN) :: VAR
```

Explanation of arguments:

AIC is a measure of model fit and parsimony, calculated as shown in Table 14-1.

BIC is a measure of model fit and parsimony, calculated as shown in Table 14-1.

DTLA is the determinant of the Fisher Information Matrix, calculated as shown in Table 14-1.

HQ is a measure of model fit and parsimony, calculated as shown in Table 14-1.

IFO is a flag indicating parameter-estimation convergence status.

    IFO = 1 indicates convergence by meeting criterion based on percent change of parameter estimates.

    IFO = 2 indicates convergence by meeting criterion based on percent change in objective function value.

    IFO = 3 indicates non-convergence.

ITERP is the number of parameter-estimation iterations performed.

KASHYAP is a measure of model fit and parsimony, calculated as shown in Table 14-1.

MLOFD is the Maximum Likelihood Objective Function evaluated for observations, calculated as shown in Table 14-1.

MLOFDP is the Maximum Likelihood Objective Function evaluated for observations and prior information, calculated as shown in Table 14-1.

MODELLENGTH is the length unit used in the model.

MODELMASS is the mass unit used in the model.

MODELNAME is the model name.

MODELTIME is the time unit used in the model.

MPR is the number of prior-information equations.

NDINC is the number of observations included in the final iteration of the parameter estimation.

NPE is the number of adjustable parameters.

NPERD is the number of parameters estimated in the iteration from which the optimal values are determined.

NPS is the total number of parameters in the parameter input blocks.

NOBS is the number of observations.

OUTNAM is the root name for output files.

STAT1 is the correlation of weighted residuals for dependent variable observations to standard normal statistics.

STAT2 is the correlation of weighted residuals for dependent variable observations and prior-information equations to standard normal statistics.

VAR is the calculated error variance.

Subroutine STA_UEV_DX_WRITE_NM

Description: This subroutine makes calls to calculate values for, and to write, the data-exchange _nm file which contains ordered, weighted residuals and their associated standard normal statistics for plotting a probability graph on a linear scale.

Argument list:

```
(MPR, NOBS, OBSNAM, OUTNAM, PLOTSYMBOLPRI, PRINAM, WTDRESIDS, WTDRESIDSPRI)
```

Declarations for argument-list variables:

```
INTEGER,                                   INTENT(IN) :: MPR
INTEGER,                                   INTENT(IN) :: NOBS
CHARACTER(LEN=LENDNAM), DIMENSION(NOBS),   INTENT(IN) :: OBSNAM
CHARACTER(LEN=MAX_STRING_LEN),             INTENT(IN) :: OUTNAM
INTEGER,                DIMENSION(MPR),    INTENT(IN) :: PLOTSYMBOLPRI
CHARACTER(LEN=LENDNAM), DIMENSION(MPR),    INTENT(IN) :: PRINAM
DOUBLE PRECISION,       DIMENSION(NOBS),   INTENT(IN) :: WTDRESIDS
DOUBLE PRECISION,       DIMENSION(MPR),    INTENT(IN) :: WTDRESIDSPRI
```

Explanation of arguments:

MPR is the number of prior-information equations.

NOBS is the number of observations.

OBSNAM contains the observation names.

OUTNAM is the root file name for output files.

PLOTSYMBOLPRI is the plot flag for the prior-information equations.

PRINAM contains the prior information names.

WTDRESIDS contains the weighted residuals.

Appendix L: Statistics (STA) Module of Chapter 14

WTDRESIDSPRI contains the weighted residuals of prior information.

Subroutine STA_UEV_FIT

Description: This subroutine calculates statistics describing the fit of simulated equivalents to observations. Results of the calculations are optionally written to a specified unit number.
Argument list:
```
(IOUT, MPR, NOBS, IPRINT, DONOTEVALRUNS, MODELVAL, MODELPRIVAL, OBSNAM, OBSVAL,
OMIT, OUTNAM, PLOTSYMBOLPRI, PRILN, PRINAM, PRIVAL, PRIWTCORR, PRIWTMATSQR,
RESIDS, RESIDSPRI, WTCORRELATED, WTDRESIDS, WTDRESIDSPRI, WTMATSQR, AVET, NNEGT,
NPOST, NRUNS, RSQ, RSQP, DNPP, NDINC, WTRL)
```

Declarations for argument-list variables:
```
INTEGER,                                        INTENT(IN)   :: IOUT
INTEGER,                                        INTENT(IN)   :: MPR
INTEGER,                                        INTENT(IN)   :: NOBS
LOGICAL, DIMENSION(3),                          INTENT(IN)   :: IPRINT
LOGICAL,                                        INTENT(IN)   :: DONOTEVALRUNS
DOUBLE PRECISION,        DIMENSION(NOBS),       INTENT(IN)   :: MODELVAL
DOUBLE PRECISION,        DIMENSION(MPR),        INTENT(IN)   :: MODELPRIVAL
CHARACTER(LEN=LENDNAM),  DIMENSION(NOBS),       INTENT(IN)   :: OBSNAM
DOUBLE PRECISION,        DIMENSION(NOBS),       INTENT(IN)   :: OBSVAL
INTEGER,                 DIMENSION(NOBS),       INTENT(IN)   :: OMIT
CHARACTER(LEN=MAX_STRING_LEN),                  INTENT(IN)   :: OUTNAM
INTEGER,                 DIMENSION(MPR),        INTENT(IN)   :: PLOTSYMBOLPRI
INTEGER,                 DIMENSION(MPR),        INTENT(IN)   :: PRILN
CHARACTER(LEN=LENDNAM),  DIMENSION(MPR),        INTENT(IN)   :: PRINAM
DOUBLE PRECISION,        DIMENSION(MPR),        INTENT(IN)   :: PRIVAL
LOGICAL,                 DIMENSION(MPR),        INTENT(IN)   :: PRIWTCORR
TYPE (CDMATRIX),                                INTENT(IN)   :: PRIWTMATSQR
DOUBLE PRECISION,        DIMENSION(NOBS),       INTENT(IN)   :: RESIDS
DOUBLE PRECISION,        DIMENSION(MPR),        INTENT(IN)   :: RESIDSPRI
LOGICAL,                 DIMENSION(NOBS),       INTENT(IN)   :: WTCORRELATED
DOUBLE PRECISION,        DIMENSION(NOBS),       INTENT(IN)   :: WTDRESIDS
DOUBLE PRECISION,        DIMENSION(MPR),        INTENT(IN)   :: WTDRESIDSPRI
TYPE (CDMATRIX),                                INTENT(IN)   :: WTMATSQR
DOUBLE PRECISION,                               INTENT(INOUT) :: AVET
INTEGER,                                        INTENT(INOUT) :: NNEGT
INTEGER,                                        INTENT(INOUT) :: NPOST
INTEGER,                                        INTENT(INOUT) :: NRUNS
DOUBLE PRECISION,                               INTENT(INOUT) :: RSQ
DOUBLE PRECISION,                               INTENT(INOUT) :: RSQP
DOUBLE PRECISION, DIMENSION(NOBS+MPR,2),        INTENT(OUT)  :: DNPP
INTEGER,                                        INTENT(OUT)  :: NDINC
DOUBLE PRECISION,                               INTENT(OUT)  :: WTRL
```

Explanation of arguments:
IOUT is the unit number associated with the file to which tabulated residuals, sum of squared residuals, and (or) summary statistics are to be written.
MPR is the number of prior-information equations.
NOBS is the number of observations.
IPRINT contains flags that control writing of various values to unit IOUT. For each element, true indicates the following values will be written:
(1) List of all observations and residuals.

262

(2) Sum of squared, weighted residuals.

(3) Summary statistics: Maximum weighted residual, minimum weighted residual, average weighted residual, number of residuals greater than zero, number of residuals less than zero, and number of runs.

DONOTEVALRUNS is a flag that, when true, indicates that the runs statistics are not to be calculated and, when false, they are to be calculated.

MODELVAL contains the simulated-equivalent values.

MODELPRIVAL contains the current values of prior information.

OBSNAM contains the observation names.

OBSVAL contains the observed values of dependents.

OMIT contains 0 if the observation is included and a negative value if it is not.

OUTNAM is the root file name for output files.

PLOTSYMBOLPRI is the plot flag for the prior-information equations.

PRILN contains a flag indicating whether the prior-information equation is log-transformed.

PRINAM contains the prior information names.

PRIVAL contains the prior information values.

PRIWTCORR is true for each prior-information equation that is correlated with one or more other prior-information equation; it is false for uncorrelated equations.

PRIWTMATSQR is the compressed square root of the correlated weight matrix on prior information.

RESIDS contains the residuals for dependents.

RESIDSPRI contains the residuals for prior information.

WTCORRELATED is true for each observation that is included in a group within which observations are correlated, and false otherwise.

WTDRESIDS contains the weighted residuals of observations.

WTDRESIDSPRI contains the weighted residuals of prior information.

WTMATSQR is the square root of the weight matrix for dependents.

AVET is used calculate average weighted residual.

NNEGT is used to accumulate number of residuals less than zero.

NPOST is used to accumulate number of residuals greater than zero.

NRUNS is the number of runs, where a run is a sequence of residuals of the same sign.

RSQ is the sum of squared weighted residuals of dependents.

RSQP is the sum of squared weighted residuals of prior information.

DNPP contains weighted residual and plot symbol for each observation, to be used for preparing probability plot data.

NDINC is the number of observations included in calculations performed in this pass through this subroutine. These observations are: (1) included in the calculation of RSQ, NNEGT, NPOST, NRUNS, and AVET; (2) written to data-exchange files, and (3) considered in populating DNPP.

WTRL is the weighted residual of the last observation for which the diagonal term of the WTMAT matrix is not less than zero.

Notes:

1. Subroutine STA_UEV_INIT should be called before calling STA_UEV_FIT to initialize the AVET, NNEGT, NPOST, NRUNS, RSQ, and RSQP variables.

Subroutine STA_UEV_INIT

Description: This subroutine initializes variables for UEV routines of the Statistics Module.

Argument list:
```
(AVET, NNEGT, NPOST, NRUNS, RSQ, RSQP)
```

Declarations for argument-list variables:
```
DOUBLE PRECISION, INTENT(OUT) :: AVET
INTEGER,          INTENT(OUT) :: NNEGT
INTEGER,          INTENT(OUT) :: NPOST
INTEGER,          INTENT(OUT) :: NRUNS
DOUBLE PRECISION, INTENT(OUT) :: RSQ
DOUBLE PRECISION, INTENT(OUT) :: RSQP
```

Explanation of arguments:
  AVET is used calculate average weighted residual.
  NNEGT is used to accumulate number of residuals less than zero.
  NPOST is used to accumulate number of residuals greater than zero.
  NRUNS is used to calculate the number of runs.
  RSQ is used to calculate the weighted sum-of-squares objective function for dependents.
  RSQP is used to calculate the weighted sum-of-squares objective function for prior information.
Note:
  NRUNS is initialized to 1; all other variables are initialized to zero.

Public Subroutines – Evaluate (EVA)

This section describes subroutines used to evaluate final results. The subroutines are presented in alphabetical order as follows:
  Subroutine STA_EVA_COMMENTS
  Subroutine STA_EVA_CRITVAL_R2N
  Subroutine STA_EVA_ORDER
  Subroutine STA_EVA_PROB_NORM_DISTRB
  Subroutine STA_EVA_RUNS

Subroutine STA_EVA_COMMENTS

Description: This subroutine writes comments on interpretation of the correlation between ordered weighted residuals and standard normal statistics. The comments are customized based on number of residuals and the value of the standard normal statistics.
Argument list:
```
(IOUT, NOBS, MPR)
```

Declarations for argument-list variables:
```
INTEGER, INTENT(IN) :: IOUT
INTEGER, INTENT(IN) :: NOBS
INTEGER, INTENT(IN) :: MPR
```

Explanation of arguments:
  IOUT is the unit number associated with the file to which output is to be written.
  NOBS is the number of observations.
  MPR is the number of prior-information equations.

Appendix L: Statistics (STA) Module of Chapter 14

Subroutine STA_EVA_CRITVAL_R2N

Description: This subroutine calculates and writes the critical values of the R2N statistic (that is, the correlation between weighted residuals and standard normal statistics) below which the hypothesis that weighted residuals are independent and normally distributed is rejected. From Shapiro and Francia (1972) and Brockwell and Davis (1987, p. 304).

Argument list:
```
(NDR, RN205, RN210)
```

Declarations for argument-list variables:
```
INTEGER,            INTENT(IN)  :: NDR
DOUBLE PRECISION, INTENT(OUT) :: RN205
DOUBLE PRECISION, INTENT(OUT) :: RN210
```

Explanation of arguments:
> NDR is the number of residuals considered.
> RN205 is the statistic for the 5% significance level.
> RN210 is the statistic for the 10% significance level.

Subroutine STA_EVA_ORDER

Description: Orders the residuals that are stored in the array D.

Argument list:
```
(MPR, NOBS, OBSNAM, PLOTSYMBOLPRI, PRINAM, WTDRESIDS, WTDRESIDSPRI, D, R,
ORDDID)
```

Declarations for argument-list variables:
```
INTEGER,                                      INTENT(IN)    :: MPR
INTEGER,                                      INTENT(IN)    :: NOBS
CHARACTER(LEN=LENDNAM), DIMENSION(NOBS),       INTENT(IN)    :: OBSNAM
INTEGER, DIMENSION(MPR),                       INTENT(IN)    :: PLOTSYMBOLPRI
CHARACTER(LEN=LENDNAM), DIMENSION(MPR),        INTENT(IN)    :: PRINAM
DOUBLE PRECISION, DIMENSION(NOBS),             INTENT(IN)    :: WTDRESIDS
DOUBLE PRECISION, DIMENSION(MPR),              INTENT(IN)    :: WTDRESIDSPRI
DOUBLE PRECISION, DIMENSION(NOBS+MPR,2),       INTENT(INOUT) :: D
DOUBLE PRECISION, DIMENSION(NOBS+MPR),         INTENT(INOUT) :: R
CHARACTER(LEN=LENDNAM), DIMENSION(NOBS+MPR),   INTENT(INOUT) :: ORDDID
```

Explanation of arguments:
> MPR is the number of prior-information equations.
> NOBS is the number of observations.
> OBSNAM contains the observation names.
> PLOTSYMBOLPRI is the plot flag for prior-information equations.
> PRINAM contains the prior-information names.
> WTDRESIDS contains the weighted residuals of dependents.
> WTDRESIDSPRI contains the weighted residuals of prior information.
> D holds the residual in the first position and the plot symbol identifier in the second position.
> R holds the standard normal statistic that correlates with the ordered residual.
> ORDDID holds the observation name in the same order as the values in D.

Subroutine STA_EVA_PROB_NORM_DISTRB

Description: This subroutine finds the probability related to a specified number of standard deviations from the mean (ip=1), or the number of standard deviations from the mean related to a specified probability (ip=-1), for a standard gaussian distribution.

Argument list:

```
(U, RNORM, IP)
```

Declarations for argument-list variables:

```
DOUBLE PRECISION, INTENT(INOUT) :: U
DOUBLE PRECISION, INTENT(INOUT) :: RNORM
INTEGER,          INTENT(IN)    :: IP
```

Explanation of arguments:

U is the associated number of standard deviations.

RNORM is the cumulative probability from a standard gaussian distribution.

IP is a flag indicating whether to find the probability related to a U (IP = 1), or a U related to a probability (IP = -1) for a standard Gaussian distribution.

Subroutine STA_EVA_RUNS

Description: This subroutine calculates and writes the runs statistic for all residuals.

Argument list:

```
(AVET, NRSO, NPOST, NNEGT, NRUNS, IOUT)
```

Declarations for argument-list variables:

```
DOUBLE PRECISION, INTENT(IN) :: AVET
INTEGER,          INTENT(IN) :: NRSO
INTEGER,          INTENT(IN) :: NPOST
INTEGER,          INTENT(IN) :: NNEGT
INTEGER,          INTENT(IN) :: NRUNS
INTEGER,          INTENT(IN) :: IOUT
```

Explanation of arguments:

AVET is used to calculate average weighted residual.

NRSO is the number of included observations.

NPOST is used to accumulate number of residuals greater than zero.

NNEGT is used to accumulate number of residuals less than zero.

NRUNS is used to calculate the number of runs.

IOUT is the unit number associated with the file to which output is to be written.

Public Subroutine – Cleanup (CLN)

Subroutine STA_CLN

Description: This subroutine deallocates all allocatable arrays that are stored in the Statistics Module.

Argument list: Empty

Index

Index

www.ingramcontent.com/pod-product-compliance
Lightning Source LLC
Chambersburg PA
CBHW081435170526
45166CB00008B/2205